GEOTECHNICAL
REPORTS FOR

SUGGESTED GUIDELINES

PREPARED BY
Technical Committee on Geotechnical Reports
of the Underground Technology Research Council

SPONSORED BY
The Construction Institute (CI)
of the American Society of Civil Engineers

American Institute of Mining, Metallurgical, and Petroleum Engineers

COMMITTEE CHAIRMAN
Randall J. Essex, P.E.

Published by the American Society of Civil Engineers

Library of Congress Cataloging-in-Publication Data

Geotechnical baseline reports for construction suggested guidelines / Prepared by Technical Committee on Geotechnical Reports of the Underground Technology Research Council ; Randall J. Essex.
　　p. cm.
　Sponsored by the Construction Institute of the American Society of Civil Engineers and the American Institute of Mining, Metallurgical and Petroleum Engineers.
　Includes bibliographical references and index.
　ISBN-13: 978-0-7844-0930-5
　ISBN-10: 0-7844-0930-7
　1. Underground construction. 2. Engineering geology--Specifications. 3. Engineering contracts. I. Essex, Randall J. II. ASCE Technical Council on Research. Underground Technology Research Council. Technical Committee on Geotechnical Reports. III. American Society of Civil Engineers. Construction Division. IV. American Institute of Mining, Metallurgical and Petroleum Engineers.

　TA712.G46　2007
　624.1'9--dc22　　　　　　　　　　　　　　　　　　　　　　　　　　　　　　　　　2007016514

American Society of Civil Engineers
1801 Alexander Bell Drive
Reston, Virginia, 20191-4400

www.pubs.asce.org

Any statements expressed in these materials are those of the individual authors and do not necessarily represent the views of ASCE, which takes no responsibility for any statement made herein. No reference made in this publication to any specific method, product, process, or service constitutes or implies an endorsement, recommendation, or warranty thereof by ASCE. The materials are for general information only and do not represent a standard of ASCE, nor are they intended as a reference in purchase specifications, contracts, regulations, statutes, or any other legal document. ASCE makes no representation or warranty of any kind, whether express or implied, concerning the accuracy, completeness, suitability, or utility of any information, apparatus, product, or process discussed in this publication, and assumes no liability therefore. This information should not be used without first securing competent advice with respect to its suitability for any general or specific application. Anyone utilizing this information assumes all liability arising from such use, including but not limited to infringement of any patent or patents.

ASCE and American Society of Civil Engineers—Registered in U.S. Patent and Trademark Office.

Photocopies and reprints.
You can obtain instant permission to photocopy ASCE publications by using ASCE's online permission service (www.pubs.asce.org/authors/RightslinkWelcomePage.html). Requests for 100 copies or more should be submitted to the Reprints Department, Publications Division, ASCE, (address above); email: permissions@asce.org. A reprint order form can be found at www.pubs.asce.org/authors/reprints.html.

Copyright © 2007 by the American Society of Civil Engineers. All Rights Reserved.
ISBN 13: 978-0-7844-0930-5
ISBN 10: 0-7844-0930-7
Manufactured in the United States of America.

Cover Photos. *TBM:* Courtesy of Ontario Hydro Generation. *Sound Transit's Beacon Hill Concourse:* Courtesy of Hatch Mott MacDonald. *Cover/Cut Sewer:* Courtesy Peter Kiewit.

16 15 14 13 12　　5 6 7 8 9

Contents

Dedication to James P. Gould		*v*
Dedication to E. B. Waggoner		*vii*
Acknowledgments		*ix*
Executive Summary		1
1.0	Introduction	4
	1.1 The Need for Review	4
	1.2 The Geotechnical Baseline Report	4
	1.3 Purpose of the GBR	6
	1.4 Purpose and Scope of This Document	6
2.0	Background	8
	2.1 Improved Contracting Practices	8
	2.2 Contractual Geotechnical Reports	9
	2.3 Shortcomings of Previous Practice	10
3.0	Geotechnical Reports	11
	3.1 Geotechnical Data Report	11
	3.2 Geotechnical Memoranda for Design	11
	3.3 Geotechnical Baseline Report	12
4.0	Differing Site Conditions Clause	13
	4.1 Historical Development	13
	4.2 Standard Clause	14
	4.3 Modifications to the Standard Clause	15
5.0	The Concept of a Baseline	16
	5.1 Baselines	16
	5.2 Contractual Assumptions	18
	5.3 Where to Set the Baseline	19
	5.4 Baseline Not a "Warranty" of Conditions to be Encountered	20
	5.5 Link with the Other Contract Documents	20
6.0	Preparation of a Geotechnical Baseline Report	22
	6.1 Organization and Content	22
	6.2 Writing the GBR – Who, When, and How	22
	6.3 Risk Registers	27
	6.4 Wording Suggestions	27
	6.5 Baseline Examples	28
	6.6 Consistency	28
	6.7 Time and Budget for Preparation	30
	6.8 Owner Involvement	31
7.0	Applications for Other Excavations and Foundations	32
	7.1 Amplification of Impacts	32
	7.2 Baselines for Small Projects	32

	7.3	Identification of Risk Factors	33
	7.4	Baseline Parameters for Consideration	34
8.0	Design-Build Procurement		37
	8.1	Site Exploration	37
	8.2	Geotechnical Data Report	37
	8.3	Geotechnical Baseline Report	38
	8.4	Recent Applications	40
9.0	Owner Perspectives		43
	9.1	Realities in the Public Sector	43
	9.2	Setting the Baseline	43
	9.3	Managing the Owner's Risk	45
10.0	Roles and Responsibilities		48
11.0	Lessons Learned		51
	11.1	GBR Preparation	51
	11.2	April 2004 Workshop	51
	11.3	June 2006 Workshop	55

List of Abbreviations   59

References   59

Index   61

Dedication to James P. Gould

The second edition of this publication is dedicated to the memory of James P. Gould. Jim is remembered by all who knew and worked with him as a warm, caring man whose good humor was equal to his engineering brilliance. He had an uncanny ability to rapidly filter engineering information, pick out the important issues, identify the problems and suggest a solution. He was a pioneer in promoting alternative means of dispute resolution in underground projects and prepared one of the first Geotechnical Design Interpretive Reports included as part of the contract documents on the Washington, DC Metro subway system.

Born in 1923, Jim grew up in Seattle and graduated with a BSCE from the University of Washington in 1944. This was followed by a hitch in the U.S. Army Corps of Engineers. After World War II, Jim studied at MIT where he received his MSCE in 1946, and at Harvard, where working with Arthur Casagrande, he earned the ScD in 1949. He then spent four years with the U.S. Bureau of Reclamation, where he worked on earth dam projects. In 1953, he jointed Moran, Proctor, Mueser and Rutledge in New York. Jim became a partner in the firm in 1973. This firm became Mueser Rutledge Consulting Engineers in 1985.

Jim's active career included a long list of difficult and noteworthy projects. Among them was geotechnical work on the U.S. Capitol, the U.S. House of Representatives, the National Gallery of Art and the Smithsonian Institution in Washington, D.C., New York's Battery Park City, and numerous port and marine projects. As a result of his work in leading Mueser Rutledge's geotechnical services during 30 years of construction of the Washington D.C. Metro Subway system, he became a valued member of advisory and consulting boards established for other major projects. Jim served on such Boards for highway and rapid transit tunnel projects in Boston, Los Angeles, Dallas and San Juan, Puerto Rico, for the Super Collider in Texas, and the Channel Tunnel between England and France.

He was a former Executive Committee chairman of the Geotechnical Division of ASCE and was active on technical committees on Earth Retaining Structures, Grouting, Tunnel Lining Design, and Groundwater. He delivered ASCE's highly esteemed Terzaghi Lecture in San Francisco in 1990, where he was named an Honorary Member. He was a member of the National Academy of Engineering, National Research Council, Transportation Research Board and the New York Academy of Sciences. He received the prestigious Moles Member Award in 1992, recognizing his outstanding contributions to the field of underground construction.

Jim's Terzaghi Lecture, published in the July 1995 Journal of Geotechnical Engineering was entitled, "Geotechnology in Dispute Resolution." That paper strongly endorsed the use of geotechnical baseline reports for underground construction by describing the need for, and importance of, a workable differing site conditions (DSC) disputes process in underground construction contracts. The paper also described the historical background of the work by the Underground Technology Research Council (UTRC) in developing a series of documents that led to industry-wide acceptance of the use of Geotechnical Baseline Reports, and the publication to which this dedication applies.

The paper emphasizes the importance of carefully planned and reported geotechnical investigations (a life long passion of Jim's) and the need for preparation and review of geotechnical baseline reports by experienced senior personnel. Jim discussed eight case histories involving DSC claims that illustrate many of the difficulties in site characterization and description that can lead to encountering unexpected conditions during construction.

Jim was an influential mentor to many of us, and is remembered fondly by virtually everyone whose life he touched.

Dedication to E. B. Waggoner

The first edition of this publication was dedicated to the memory of Eugene B. Waggoner. Gene is remembered by those who knew him as a caring, warm man, who always had a humorous story relevant to the situation at hand. His ability to enhance a story or joke over time was legendary. As a professional, he was one of the pioneers who developed the classic study of geology into the applied field we recognize today as engineering geology. He understood geologic processes, structural geology, mineralogy, and petrology, and translated that understanding into solutions to difficult engineering and construction problems. Gene was one of the principal movers in the early days of interpretive geotechnical reports for construction. His vision and desire for successful construction projects through improved communication and cooperation has inspired the development of this document.

Born in 1913 in Missouri, Gene's family moved to Los Angeles during his early years. He received his BA and MA in geology from UCLA and set out in 1939 to become a petroleum geologist. He joined the U.S. Bureau of Reclamation in 1942, where he changed his focus to civil engineering projects. He left the government in 1954 and went into private practice, and in 1960, merged his practice with Woodward-Clyde Associates. Gene became president of Woodward-Clyde in 1967, and retired in 1973 to have open-heart surgery. His "retirement" was more active than most professionals' careers during their productive years. He passed away in 1991 after a bout with pneumonia.

Gene was an internationally recognized expert in engineering geology. He worked on hundreds of major projects in more than 50 countries, largely in the fields of dam and underground construction. Those who sought his expertise included the U.S. Department of State, Defense Intelligence Agency, World Bank, United Nations, the U.S. Corps of Engineers, Federal Energy Regulatory Commission, Royal Irrigation Department of Thailand, Greek Ministry of Public Works, Swaziland Electric Board, and many worldwide engineering firms, U.S. earthworks construction contractors, and attorneys practicing in the construction contracts field.

Gene was a member of the National Academy of Engineering, the U.S. National Committee on Tunneling Technology, the Geotechnical Board of the National Research Council, was an honorary Member of the Association of Engineering Geologists, President of the American Consulting Engineers Council, a Fellow of the American Society of Foundation Engineers, and Life Member of the American Society of Civil Engineers.

In the early 1970s he worked on two reports, and in the 1980s led the preparation of a third report for the National Research Council, which served to change the course of underground construction in the U.S. These reports, *Better Contracting for Underground Construction, Better Management of Major Underground Construction Projects,* and *Geotechnical Site Investigations for Underground Projects,* are foundations for many of the concepts discussed in this document.

This dedication should end as it began, by emphasizing Gene's humanity. He had integrity, intelligence, humility, humor, and self-confidence. His greatest asset was that he loved people, and always made the effort to connect with them whether they were laborers with a third grade education or the Royal Family of Thailand. All were treated with respect and dignity. The heavy and underground construction industry will miss Gene's professional insight and dedication to his industry and his country, as well as his good will.

Acknowledgments

This document is the product of many practitioners' contributions over a period of more than 16 years. The first edition evolved from a number of contributors over a period of three years. The second edition was prepared by a working group who offered revisions, updates, and new text to reflect updates and lessons learned. The principal author of this document is Randall J. Essex, Chairman of the Technical Committee for Geotechnical Reports, Underground Technology Research Council.

The following individuals comprised the Technical Committee assembled in 2006 who contributed to this second edition:

Randall Essex, Chairman
Hatch Mott MacDonald
Rockville, Maryland

S.H. "Bart" Bartholomew
Consultant
Chico, California

Peter Douglass
Consultant
Seattle, Washington

Robert Fitzgerald
Watt, Tieder, Hoffar & Fitzgerald
McLean, Virginia

Ronald Heuer
Consultant
McHenry, Illinois

Stephen Klein
Jacobs Associates
San Francisco, California

Daniel Meyer
Consultant
Lake Forest, Illinois

James Monsees
Parsons Brinckerhoff, Quade and Douglass
Orange, California

James Morrison
Kiewit Engineering Company
Omaha, Nebraska

Robert Pond
Frontier-Kemper
Evansville, Indiana

P.E. "Joe" Sperry
Consultant
Auburn, California

Richard Switalski
Northeast Ohio Regional Sewerage District
Cleveland, Ohio

The following individuals comprised the Technical Committee assembled in 1993 who contributed to the first edition:

Randall Essex
Tor Brekke
Peter Douglass
Larry Heflin
Roger Ilsley

Al Mathews
Terrence McCusker
James Monsees
Red Robinson
Robert Smith

Text for the first edition was adapted from position papers presented during an industry forum held in 1996. The papers were prepared by panels chaired by the following individuals:

Contractor's Perspective:
Robert Pond

Consultant's Perspective:
Gregg Korbin

Owner's Perspective:
Walter Mergelsberg

DRB Member's Perspective:
P.E. "Joe" Sperry

Designer's Perspective:
James Monsees

Legal Perspective:
Robert Smith

EXECUTIVE SUMMARY

Background

Since the 1970s several forms of interpretive geotechnical reports have been incorporated into the Contract Documents for underground construction projects. Sixteen years ago a number of practitioners felt the need to re-examine the role of this type of report and its benefit in contracting. Poorly written and ambiguous interpretive geotechnical reports, and inconsistencies between the interpretive report and other Contract Documents, were doing more harm than good in the effort to avoid and resolve construction disputes.

The first edition of this document was published in 1997 based on feedback obtained from three industry forums conducted between 1993 and 1996. In 2004, as additional experience was gained using Geotechnical Baseline Reports, and their application expanded into other types of subsurface construction and design-build construction, the need for a second edition became apparent. This second edition is based on feedback obtained during industry forums in 2004 and 2006.

This guideline is intended to serve as a reference for preparers and users of GBRs, and to inform Owners of the importance of the contents of the GBR in the allocation of financial risk. This guideline focuses on subsurface projects involving tunnels and shafts, but also expands the applicability of GBRs to deep foundations, open-cut pipelines, braced and tied-back excavations, and highways.

Though the information contained in this document represents a consensus opinion within the industry on a range of issues, the opinions of practitioners vary on a number of topics. The suggestions provided in this document are therefore intended as guidelines, and should not be interpreted as rules, requirements, or standards of care.

Recommendations

It is recommended that a single interpretive report be included in the Contract Documents and be called a *Geotechnical Baseline Report (GBR)*. The primary purpose of the GBR is to establish a single source document where contractual statements describe the geotechnical conditions anticipated (or to be assumed) to be encountered during underground and subsurface construction. The contractual statement(s) are referred to as baselines. Risks associated with conditions consistent with or less adverse than the baselines are allocated to the Contractor, and those materially more adverse than the baselines are accepted by the Owner. Other important objectives of the GBR are to discuss the geotechnical and site conditions related to the anticipated means and methods of constructing the underground elements of the project.

The factual information gathered during the project investigations should be summarized in a Geotechnical Data Report (GDR). The GDR should be included as a Contract

Document, however the GBR should be clearly indicated as taking precedence over the GDR within the Contract Documents hierarchy.

Other interpretive reports may be prepared by the design team, addressing a broad range of design issues for the team's internal consideration. Such reports should be referred to as Geotechnical Design Memoranda, and should be clearly differentiated from the GBR. The GBR should be the only interpretive report prepared for use in bidding and constructing the project. Preparation of other interpretive reports in the course of the final design, such as a Geotechnical Interpretive Report (GIR), is superfluous, a potential source of confusion and conflict, and is strongly discouraged.

A Differing Site Conditions (DSC) clause is almost always included as a standard in the general conditions or general provisions of a contract that will involve subsurface construction. Continuation of this practice is strongly recommended. The DSC clause relieves the Contractor of assuming the risk of encountering conditions differing materially from those indicated or ordinarily encountered, and provides a remedy under the construction contract so that the matter can be handled as an item of contract administration. Clear, precise, and quantifiable baselines enhance the benefits and use of the DSC clause.

Baseline statements in the GBR should be interpretations expressed as contractual representations of anticipated subsurface conditions. The baselines should be meaningful, reasonable and realistic, and to the maximum extent possible should be consistent with available factual information contained in the GDR. However, if factual data is not available, or is considered to be misleading and not representative of field conditions, baselines may be based on other information (e.g., previous tunneling experience in similar geology) and engineering judgment, provided the reasoning is clearly explained. If baselines are extreme and unrealistic, the "law of unintended consequences" is likely to prevail, and the practical value of the GBR is otherwise frustrated.

Owners usually retain design consultants (civil and geotechnical engineers, geologists, and hydrogeologists/hydrologists) who are familiar with the local geology, have design and construction experience with similar projects, or both. The Owner should ensure that these individuals are intimately involved, in a collaborative manner, with the preparation and review of the GBR.

Owners should participate in and contribute to the setting of the baselines and should understand the consequences of the levels at which the baselines are set. The Owner may decide to allocate certain risks and costs of potential differing site conditions to the Contractor through the use of more adverse baselines. This will usually result in an increased bid price. Alternatively, the Owner may choose to share the risks and costs through less adverse baselines and utilization of either alternative payment provisions or a change order process, if the more adverse conditions materialize. In this instance, such an Owner may enjoy a lower bid price but in either case, the costs associated with handling the site conditions are the Owner's responsibility. The difference is that in the latter case, the Owner will accrue certain financial benefits if the adverse conditions are not realized.

This is true whether the project utilizes a traditional design-bid-build or design-build procurement approach.

For design-bid-build procurement, GBR preparation should begin only after the design has been advanced to a certain level of completion (usually the 50 or 60 percent level as a minimum). GBR preparation should be a collaborative effort among representatives of the design team, including the project geotechnical consultant and the project Owner.

For design-build procurement, including public-private partnerships, a modified three-step approach to GBR preparation is recommended where the Owner's design team prepares an initial document (Step 1) that provides a common basis for bidders. Each design-build team then supplements the report with their design-based considerations (Step 2). Only upon the Owner's review, clarification, and interaction with a preferred design-build team (Step 3) is the final GBR version vetted and ratified.

The document presented herein contains recommended guidelines for what should and should not be included in the GBR, provides a checklist of items to consider, provides recommendations for the content and wording of baseline statements that will improve their clarity, understanding and usefulness, and presents examples of problematic and improved practice in stating baselines. This document also discusses the importance and benefit of ensuring compatibility between the GBR and other elements of the Contract Documents.

Focus of Second Edition

This second edition provides new perspectives in the following areas:

- applications in addition to tunnel and shaft construction, such as deep foundations, open cut pipelines, braced and tied-back excavations, and highways; and
- application of GBRs to design-build procurement.

Perhaps most importantly, this second edition contains a chapter that summarizes lessons learned from the experiences of Owners, Contractors, designers, consultants, and dispute adjudicators through application of the initial guidelines document between 1997 and 2006. The application of GBRs, along with other contracting practices that have similarly evolved, can be improved upon if the industry captures the experiences of practitioners and presents them in guideline documents such as this one. The industry's challenge is to appreciate those lessons, and take them to heart when scoping, writing, reviewing, and interpreting GBRs on subsequent projects.

1.0 INTRODUCTION

1.1 The Need for Review

The underground construction industry in North America has made significant advances in the development and implementation of methods to avoid and resolve disputes during construction. This evolved in the wake of years of disruptive and antagonistic project disputes, and ensuing time consuming and costly litigation. In the early years, problems developed with the interpretive geotechnical report that was often included as a Contract Document. Some felt the need to re-examine its role and benefit in the contracting process. Others felt that poorly written, overly general, and ambiguous interpretive geotechnical reports, and inconsistencies between interpretive reports and other Contract Documents, were doing more harm than good.

The Technical Committee on Geotechnical Reports of the Underground Technology Research Council (UTRC) obtained feedback from the industry on the subject of interpretive geotechnical reports. The first edition of this publication in 1997 was based on input obtained from more than 150 individuals in the industry in three forums held between 1993 and 1996. The focus of the first edition reflected industry consensus that while interpretive geotechnical reports play a significant role in avoiding and resolving disputes in underground construction, their content, presentation of baselines, and consistency with other contract documents could be improved substantially.

This second edition incorporates additional perspectives and lessons learned from two industry forums held in 2004 and 2006. The final manuscript was reviewed by the UTRC Technical Committee on Geotechnical Reports which is jointly sponsored by the American Society of Civil Engineers (ASCE) and the Society of Mining Engineers, by representatives from the Geotechnical Institute of ASCE, and by the Executive Committees of the UTRC and the Construction Institute of ASCE.

The information contained in this document represents a consensus of opinion among many practitioners who have written, applied, and interpreted GBRs on significant underground construction projects over the last 20 years. Nevertheless, the workshops made clear that different opinions remain on a number of topics. The suggestions in this document should therefore be considered as guidelines and not interpreted as rules, requirements, or as defining a standard of care or code of practice.

1.2 The Geotechnical Baseline Report

For many years geotechnical data gathered during the design phase were made available to bidders, but the Contract Documents typically disclaimed any responsibility for the risk of interpreting the data. The bidders had neither the time nor the guaranteed return on investment to undertake such costly investigations and thus were forced to rely on their own interpretations of the available data. In such circumstances, bidders occasionally failed to recognize a likely problem whose identification required a substantial degree of geotechnical experience. Furthermore, in a highly competitive bidding environment, low bidders would frequently make overly optimistic interpretations of the data, arguing that they were entitled to their interpretation if reasonable/plausible. Significant claims for additional compensation were submitted when actual conditions more adverse than the bidders' interpretations were encountered.

Eventually, many Owners recognized the need to present the designer's interpretations as part of the Contract Documents. Various names were given to this report through the years. For reasons explained in Chapter 5 of this document, it is recommended that this report be called a ***Geotechnical Baseline Report (GBR)***.

Projects involving subsurface excavation present many risks, all of which must be assumed by either the Owner or the Contractor. The greatest risks are associated with the materials encountered and their behavior during excavation and installation of support. The main purpose of the GBR is to clearly define and allocate these risks between the contracting parties.

The GBR establishes a contractual understanding (interpretation) of the subsurface site conditions, referred to as baselines. Risks associated with conditions consistent with or less adverse than the baselines are allocated to the Contractor, and those significantly more adverse than the baselines are accepted by the Owner. The latter conclusion derives from the philosophy that the Owner owns the ground, as well as any obstructions in the ground. If conditions are determined to be more adverse than portrayed in the baselines, the Owner pays any additional cost of overcoming those conditions.

The manner in which a GBR is developed and presented will be different for projects procured under traditional Design-Bid-Build and Design-Build (DB) procurement. Under traditional project procurement, the GBR is used by:

- the entity preparing the technical specifications in cases when means and methods of construction are specified;

- the design team, as a basis for preparing a construction cost estimate, including contingencies, for the Owner's budgeting purposes;

- the bidders, for contractual statements of anticipated subsurface conditions and geotechnical risks allocated to the Contractor;

- the Contractor for the selection of construction means, methods and equipment;

- the Contractor and the Owner during construction, for comparing encountered subsurface conditions with the contractual baseline interpretation as the basis for identifying differing site conditions; and

- dispute adjudicators for resolution of disputes related to encountered conditions that are asserted to be more adverse than those indicated in the GBR.

With DB project procurement, GBR development may be modified as discussed in Chapter 8. However, the resulting GBR is used in the same manner as indicated above.

1.3 Purpose of the GBR

The principal purpose of the GBR is to set clear realistic baselines for conditions anticipated to be encountered during subsurface construction, and thereby provide all bidders with a single contractual interpretation that can be relied upon in preparing their bids. Other key objectives of the GBR include:

- presentation of the geotechnical and construction considerations that formed the basis of design for the subsurface components and for specific requirements that may be included in the specifications;

- enhancement of the Contractor's understanding of the key project constraints, and important requirements in the contract plans and specifications that need to be identified and addressed during bid preparation and construction;

- assistance to the Contractor or DB team in evaluating the requirements for excavating and supporting the ground; and

- guidance to the Owner in administering the contract and monitoring performance during construction.

The GBR is more than a collection of baselines. This report is the primary contractual interpretation of subsurface conditions and the report should discuss these conditions in enough detail to accurately communicate these conditions to the bidders. As noted in subsequent chapters herein, the discussions should explain the rationale for the baselines.

The GBR allocates risks depending upon how the baselines are defined. It is also a risk management tool, because it can address the resolution of circumstances beyond the baselines.

1.4 Purpose and Scope of This Document

This document is intended to serve as a guide for preparers and users of GBRs, and to inform Owners of the importance of the contents of the GBR in the allocation of financial risk. Benefits of implementing these guidelines should result in:

- Owners selecting consultants to prepare and review GBRs on the basis of demonstrated qualifications and experience in the preparation and review of such documents;

- Designers preparing more precise, clear, and quantifiable baselines;
- GBRs becoming more standard and consistent with respect to content;
- Owners having a better understanding of how the financial risks for subsurface conditions have been allocated and how they can manage their risk in a proactive manner;
- bidders having a better basis for assessing their risks and pricing the work;
- fewer disagreements among the Contractor, Owner, and Designer regarding the anticipated surface and subsurface conditions; and
- clearly defined bases for Dispute Adjudicators to determine if an asserted differing site condition has, in fact, been encountered.

Guidelines provided in this document address:

- types of information to be included in a GBR;
- types of information that should not be included in a GBR, but are more appropriately addressed elsewhere in the Contract Documents or in design-phase memoranda;
- how the GBR should be coordinated with other Contract Documents;
- wording suggestions in preparing baseline statements;
- how GBRs can be applied to a range of projects involving subsurface excavation;
- how GBRs can be developed and implemented for DB procurement; and
- key roles and responsibilities in the preparation and application of GBRs.

This Second Edition includes three new chapters:

- Chapter 7 on the application of GBRs to projects other than tunnel and shaft construction, such as deep foundations, pipelines, braced or tied-back excavations, and highway earthworks;
- Chapter 8 on the application of GBRs to DB procurement; and
- Chapter 11 that discusses lessons learned through application of the initial guidelines document between 1997 and 2006.

2.0 BACKGROUND

2.1 Improved Contracting Practices

In 1972, the Washington Metropolitan Area Transit Authority recognized the importance of describing the subsurface conditions that bidders should anticipate when preparing their bids for tunnel contracts on the Washington, D.C. Metro. It was decided to address these conditions in a separate report made a part of the Contract Documents.

During this same period, important reference documents were being developed within the construction industry to address the rising costs of underground construction, and ways to reverse the trends. The first of these reports was published in 1974 by the U. S. National Committee on Tunneling Technology (USNCTT), within the National Research Council. The report, entitled *Better Contracting for Underground Construction,* had a profound, positive influence on the tunneling industry. This document identified the fundamental need to improve the overall approach to contracting for underground construction projects, with statements such as the following:

> "...if all bidders can base their estimates on a well defined set of site conditions with assurance that equitable reimbursement will be made when changed conditions are encountered, the Owner will receive the lowest reasonable bids with a minimum of contingency for unknowns."

In the 1970s and 1980s, the old way of resolving claims in court continued to flourish, and more tunneling practitioners began to recognize the need to change their ways. In 1984, the USNCTT published a two-volume report entitled *Geotechnical Site Investigations for Underground Projects.* The report, which based its conclusions and recommendations on the partial review of 200 heavy construction projects, and a thorough review of 87 of those projects, made two fundamental contributions to the industry. First, the report demonstrated that the greater the investment in exploring, clearly communicating, and disclosing the subsurface conditions, the lower the final cost of the project. Second, the report presented a recommended outline for interpretive geotechnical reports and a checklist of items to be addressed.

In 1989, the UTRC's Technical Committee on Better Contracting Practices published a booklet entitled *Avoiding and Resolving Disputes in Underground Construction.* An updated edition was published in 1991, entitled *Avoiding and Resolving Disputes During Construction.* Both editions contained a section that discussed the objectives and contents of interpretive geotechnical reports.

2.2 Contractual Geotechnical Reports

The approach to the preparation of contractual geotechnical reports for underground construction evolved over the past 30 years. Historically, some practitioners prepared only one report, essentially a Geotechnical Data Report (GDR), which presented only factual information such as boring logs and the findings from field and laboratory tests. Interpretations and predictions as to the behavior of the indicated subsurface materials during construction were left to the bidders. Other practitioners included their interpretations in the Contract Documents, either in a report separate from the GDR, or combined with the data in a single document.

Interpretations are needed for design and for construction. At the earliest stages of the design process, geotechnical information must be reviewed to identify subsurface conditions warranting special design considerations, and to evaluate construction methods most suitable to the anticipated conditions. Because some of the options considered might be discarded later during the design, it is necessary to distinguish between interpretations addressed by the design team during the design process, and interpretations that relate specifically to the design and construction methods addressed in the Contract Documents.

A further source of variability relates to the manner and degree in which these various geotechnical reports are presented in the Contract Documents. Contract Documents are intended to define and control the construction of the work. Documents provided for information only are subsidiary to the Contract Documents, but are intended to serve as background information relevant to the project. Generally the provision of documents "for information only" has been driven by the need to make a full disclosure of all related geotechnical information, and to avoid the appearance of withholding this information from prospective bidders.

The purpose of including an interpretive geotechnical report in the Contract Documents has changed somewhat through the years. Initially, the objective was to assist Contractors in developing their own interpretations of the factual information, rather than only providing them the "facts". In providing this interpretation, it was considered appropriate to frame these interpretations within the context of the design and the designer's intent. The term *Geotechnical Design Summary Report*, as described in previous guideline documents, was intended to set forth the designer's interpretations of the anticipated subsurface conditions, and their impact on design and construction. Occasionally, when explaining the basis for design, practitioners described the uncertainties involved, and used generalized terms in their discussion that may have been geologically correct, but were ambiguous when considered as a "baseline". This ambiguity in turn led to disputes.

The critical issue is how the Contractor addresses the anticipated conditions. The GBR should have construction issues as its main focus; the basis for design, which may be addressed, should be secondary. This establishes a clear focus for why the report is prepared, how it is to be used, and how it should be written.

In the 1990s, it was suggested that the interpretive geotechnical report to be included in the Contract Documents be called a Geotechnical Baseline Report. In 1997, as follow-on to the 1991 publication, the UTRC's Technical Committee on Geotechnical Reports published the booklet entitled *Geotechnical Baseline Reports for Underground Construction,* setting out guidelines and practices for the preparation of such reports.

2.3 Shortcomings of Previous Practice

Contractual geotechnical interpretive reports that were used before the advent of GBRs had a number of shortcomings. Some of these shortcomings persist today:

- baselines, if provided, have not adequately described the conditions to be expected;
- descriptions of anticipated conditions have often been overly broad, ambiguous or qualitative, resulting in disputes over what is indicated in the Contract;
- descriptions of assumed conditions and behaviors have been much more adverse than indicated by the data, or have appeared arbitrary and unrealistic, without adequate explanation or justification for such conservatism;
- discussions have been either unnecessarily repeated or in direct conflict with information contained on the drawings, in the specifications, or other provisions of the Contract; and
- the effects of means and methods of construction excavation and support on ground behavior have not been well described.

This publication provides guidelines intended to improve the clarity of content within the GBR, and to improve the consistency between the baseline document and the other Contract Documents.

Other shortcomings had to do with two fundamental considerations on the part of some Owners who were concerned by the number and dollar value of claims by Contractors for differing site conditions, despite the existence of baseline statements in the Contract. First, these Owners often feel that the authors of the GBR should have "gotten it right at the beginning so that we are not stuck with this nasty surprise", i.e., they misunderstand the relationship of the level of risk they are, or are not, taking and the level of conservatism reflected in the baseline statements. Second, they may not understand that the level of reliability and accuracy of baseline statements is closely related to the thoroughness of the geotechnical investigations, which in turn is dependent on the time and cost the Owner is willing to invest. These sentiments led to the conclusion that additional effort is required to educate Owners about the role of baseline statements in identifying and allocating risk between the parties.

The Owner must recognize that the contractual baselines represent one interpretation of the subsurface conditions, as developed from the available information. More than one interpretation of subsurface conditions may be reasonable based on the information available during preparation of the Contract Documents. Additional investigations, involving additional time and money during the design phase, may be required to improve the level of certainty of the baseline interpretations.

3.0 GEOTECHNICAL REPORTS

3.1 Geotechnical Data Report (GDR)

The GDR is a document developed by the Designer and/or the Designer's geotechnical engineer, which contains the factual information that has been gathered during the exploration and design phases of the Project.

The GDR should contain the following information:

- a description of the geologic setting;
- a description/discussion of the site exploration program;
- the logs of all borings, trenches, and other site investigations;
- a description/discussion of all field and laboratory test programs; and
- the results of all field and laboratory testing.

The GDR must be included as a Contract Document. The GBR, in the event of conflict or ambiguity, must be given precedence over the GDR within the Contract Document hierarchy.

In the event that the GBR is silent on a particular circumstance, the GDR should be reviewed to see if there is any data/information relevant to the issue in question. This is discussed further in Chapter 11.

3.2 Geotechnical Memoranda for Design

The project design may be carried out by a multi-firm team. An interpretation of the available geologic data is often needed within the design team well in advance of the preparation of a GBR.

Following completion of site exploration activities and preparation of a draft GDR, the geotechnical firm (or designer, if the same firm) may prepare a draft memorandum for design that addresses a broad range of issues for the project team's internal consideration. The interpretive report for design may be used to:

- comment on and discuss the data;
- present one or more initial interpretations of the data;
- evaluate the limitations of the data and discuss additional data needs;
- present an evaluation of how the subsurface conditions would affect alternative approaches to project design and construction;
- evaluate project risks as a function of alternative construction approaches;

.ruction impacts on adjacent facilities; and

.echnical design criteria for both permanent and temporary subsurface

.sions may appropriately address broad ranges of anticipated conditions to ina... .ne level of certainty (or uncertainty) in these judgments. Such discussions are not appropriate as baselines. The report may discuss design and construction alternatives that are subsequently judged of unacceptably high risk to the Owner (or third parties), and are thus eliminated from further consideration and not addressed in the GBR. Because of the differences between this preliminary interpretive report and the GBR, it is recommended that a title be given to the report (or reports) that clearly portrays its intent and timing within the design process, e.g., "Draft Geotechnical Memorandum", or "Draft Geotechnical Memorandum for Design". Although the document must be disclosed to bidders as available information, it should not be a part of the Contract Documents. The report should include specific introductory statements that it is a preliminary document not to be used for construction purposes and that interpretations and discussions presented therein will be superseded by subsequent interpretations and baselines in the GBR.

Depending on the design approach and the number of design iterations that occur during the design process, multiple geotechnical memoranda, or amended or revised versions of the memoranda, may be produced. However, the GBR should be the only interpretive report that is included in the Contract Documents. Preparation of another interpretive report by the geotechnical consultant or design team in the course of the final design, such as a Geotechnical Interpretive Report (GIR), is superfluous, a potential source of confusion and conflict, and is strongly discouraged.

3.3 Geotechnical Baseline Report

The GBR should be the sole geotechnical interpretive document upon which the Contractor may rely. The GBR should be limited to interpretive discussion and baseline statements, and should make reference to, rather than repeat or paraphrase, information contained in the GDR, drawings, or specifications. Chapters 5 and 6 contain further discussion of the suggested content and format of GBRs.

4.0 DIFFERING SITE CONDITIONS CLAUSE

4.1 Historical Development

A primary purpose for the baseline statements in the GBR is to assist in the administration of the Differing Site Conditions (DSC) clause. In order to appreciate the role that the GBR serves in this regard, it is helpful to review the history of the clause. The first standardized "changed conditions" clause was developed by an Interdepartmental Board of Contracts and Adjustments on November 22, 1921 by the U.S. Bureau of the Budget. The purpose of this clause was to provide a contractual basis for relief to the Contractor for encountered site conditions that were more adverse than those indicated in the construction contract. This clause was included in a standard form of general conditions for construction contracts that was issued on August 20, 1926, and was subsequently approved by the President of the United States for use by the federal government (Mathews, 1985). To this day, Federal Regulations mandate its use in U.S. Government contracts.

The federal clause (reproduced in Section 4.2 below), or a similar provision, has been incorporated into the standard contract documents sponsored by a number of professional and public organizations, including:

- the American Institute of Architects;

- the Engineers Joint Contract Documents Committee (American Consulting Engineers Council, American Society of Civil Engineers, National Society of Professional Engineers);

- the American Society of Civil Engineers, in collaboration with the Associated General Contractors of America;

- the American Association of State Highway and Transportation Officials; and

- numerous state and local governments.

Over its 80-year history, the wording of the clause has undergone minor refinement, but the underlying principle has remained. In 1968, for example, the term "Changed Conditions" was changed to "Differing Site Conditions".

The DSC clause was developed to take at least some of the gamble on subsurface conditions out of the bidding process, and thereby reduce the bid prices. Without relief under the DSC clause, the Owner would assign all risk to the Contractor, and would thus pay all of the Contractor's contingency costs for adverse conditions, whether the adverse conditions were encountered or not. The DSC clause was developed to avoid these unnecessary costs and remove part of the risk from the Contractor.

Despite the early institution of the DSC clause, arguments continued to develop as to the conditions indicated in the Contract. Contracts typically included disclaimers that bidders should not rely on boring logs and other information obtained during the design, and encouraged Contractors to make their own subsurface investigations during the bidding phase of the project. The favorite expression among Owners and engineers was: "You bid it - you build it." Contractors responded by pursuing, and often winning, construction claims, despite the existence of the disclaimers and exculpatory clauses. Judges and juries believed that if geotechnical information were made available to bidders, they had the right to rely on this information, even when the information was disclaimed and not included in the Contract Documents. Driven by increasingly frequent litigation and escalating costs for construction, the heavy construction industry was motivated to change its approach to disputes resolution, and to provide tools to supplement the DSC clause.

4.2 Standard Clause

A DSC clause is nearly always included as a standard clause in the general conditions or general provisions of a contract that will involve subsurface construction. The Federal clause is typical. Articles (a) and (b) of the Federal clause are presented below:

DIFFERING SITE CONDITIONS (APRIL 1984)

(a) The Contractor shall promptly, and before such conditions are disturbed, give a written notice to the Contracting Officer of (1) subsurface or latent physical conditions at the site which differ materially from those indicated in this contract, or (2) unknown physical conditions at the site, of an unusual nature, which differ materially from those ordinarily encountered and generally recognized as inhering in work of the character provided for in the contract.

(b)The Contracting Officer shall investigate the site conditions promptly after receiving the notice. If the conditions do materially so differ and cause an increase or decrease in the Contractor's cost of, or time required for, performing any part of the work under this contract, whether or not changed as a result of the conditions, an equitable adjustment shall be made under this clause and the contract modified in writing accordingly.

The function of the DSC clause is twofold. First, it relieves the Contractor of assuming the risk of encountering conditions differing materially (i.e., in a significant, meaningful way) from those indicated or ordinarily encountered. Second, it provides a remedy under the construction contract, to handle the matter as an item of contract administration.

The ease of administering the DSC clause during construction depends on how well the anticipated conditions are defined. There can be more than one plausible interpretation of the subsurface data collected from the investigations. The GBR presents one such set of interpretations as the anticipated conditions under the Contract. The more clearly defined those anticipated conditions, the more easily the encountered conditions can be evaluated as being materially different or not. Clear, precise baselines enhance the benefits and use of the DSC clause, because they provide a more straightforward basis for its administration.

4.3 Modifications to the Standard Clause

Over the years, an impressive body of case law has emerged with respect to the application and interpretation of the Federal DSC clause. The Federal Clause has been modified in some jurisdictions. The authors of the GBR should understand the contractual significance of such modifications, and adjust the wording in the GBR accordingly.

Some Owners and engineers have expressed that the Owner should be entitled to a credit from the Contractor if the subsurface conditions encountered are less adverse that indicated by baselines. As the Federal DSC clause is written, only the Contractor can initiate a claim under the clause. However, once the Contractor brings forward a claim under the DSC clause, the Owner could possibly find a basis for a lowering of the Contract Price. Some believe that the standard clause should be modified to permit the Owner to initiate a claim for a credit under the clause. For the following reasons, this is not recommended.

As discussed throughout this document, the concept of a GBR is to establish baselines for contractual purposes during the performance of the construction contract. The bidders are not mandated to base their bids on the baselines stated. To the contrary, bidders may bid consistent with the baselines or less adverse than the baselines in an effort to win the work. If Contractors believe that the actual conditions will be more favorable than stated in the baselines and wish to be as competitive as possible, they may choose to accept the risk of being mistaken and bid below the stated baselines. The economic benefit of a Contractor's decision to bid below the baselines is conveyed to the Owner in a lower bid, thereby already reflecting the economic consequences of better conditions in their bid.

If the Contract Documents included a DSC clause that entitled the Owner to a downward adjustment in the contract price for encountered conditions less adverse than indicated by the baselines, bidding Contractors would have no incentive to bid below the baseline. To do so would expose them to the risk of a downward adjustment when they have already reflected such an adjustment in their bid. The lack of incentive for Contractors to bid as competitively as possible would tend to increase the bid prices, and at the very least, lead to debates as to what the Contractor did or did not assume in his bid. The use of a deductive DSC clause is not recommended.

5.0 THE CONCEPT OF A BASELINE

5.1 Baselines

The planning, design, and construction of underground projects must cope with uncertain subsurface conditions. "Mother Nature" did not create subsurface conditions in accordance with a materials properties handbook, nor do geologists or geotechnical engineers (or any other participants in the process) have magical predictive powers. The design and construction process must account for the variability of subsurface conditions, and for potential project costs associated with that variability. To establish realistic contractual baselines (not necessarily based on the most optimistic of reasonable interpretations), and have provisions to address conditions more adverse than those baseline conditions, is a reasonable and effective approach to risk allocation and acceptance.

The cost of constructing the project along a predetermined, linear path or within a limited area represents the greatest single risk associated with underground projects. The path may be optimized to a degree, but it will more often be constrained by functional, right-of-way, environmental, and constructability considerations. The geotechnical challenges presented to the design team are two-fold. One challenge is to understand the range of possible ground and groundwater conditions at the site, so that the design and the contract provisions account for those conditions. The other challenge is to realistically describe the anticipated conditions so that the financial risks of coping with them are clearly allocated between the Owner and the Contractor.

The first challenge has room for uncertainty and generality. So long as the facility can be constructed and operated under the most adverse range of conditions anticipated, it will fulfill its intended long-term function. In many cases, the variability of the subsurface conditions may have little to do with the feasibility of constructing the facility (e.g., the strength of rock formations to be bored by a tunnel boring machine), but usually will influence the cost and schedule. However, the second challenge has no room for uncertainty or generality. The less clearly the anticipated geotechnical conditions are described in the form of baseline statements, the more likely the potential for misunderstandings during construction, for disputes, and for increases in cost.

The goal of baselines is to translate the results of geotechnical investigations and previous experience into clear descriptions of anticipated subsurface conditions upon which bidders may rely. The baselines also provide the Owner with the opportunity to allocate risks associated with these conditions. Items to be addressed in baselines include:

- the estimated amounts and distribution of different materials on the project;
- description, strength, permeability, grain size, and mineralogy of the various intact materials;
- description, strength, and permeability of the ground mass as a whole;

- quality of rock mass and characteristics of discontinuities, including roughness, infilling materials and alteration;
- groundwater levels and groundwater conditions anticipated, including items such as inflows, estimated pumping volumes and rates, and anticipated groundwater chemistry;
- the anticipated behavior of the ground, and the impact of groundwater, with regard to applicable methods of excavation and installation of ground support;
- construction impacts on adjacent facilities;
- potential or known faults, shears, fault zones, and shear zones; and
- other geotechnical and known man-made sources of potential difficulty or hazard that could impact the construction process, such as boulders, abandoned piles, buried utilities, buried debris and other obstructions, high or low top of bedrock, mixed face conditions, geologic contacts, gas, and contaminated ground and groundwater.

To the maximum extent possible, baseline statements are best described using quantitative terms that can be measured and verified during construction. The importance of this point cannot be overstated. Qualitative descriptions, if required, should be clearly defined using generally accepted industry definitions such as those published by the American Society for Testing and Materials (ASTM), International Society for Rock Mechanics (ISRM), American Society of Civil Engineers (ASCE), and other recognized standards.

However, some baseline issues may be qualitative, and not definable in quantitative, measurable terms. For example, an appropriate baseline statement might be:

> "Even though the P2 sand has high SPT N values, when exposed in an open face below the groundwater table, it will tend to exhibit unstable, flowing ground behavior unless positive measures are taken to prevent the flowing behavior from developing."

In other instances, baselines may be appropriately stated in qualitative terms, but may not be reliably measured during construction. For example, the occurrence of boulders might be described in terms of the number of boulders in different size ranges, as a percentage of the excavated volume, or as a certain number of boulders of an assumed excavation dimension. Shaft excavation might facilitate the detection and counting of encountered boulders. But in a TBM-mined tunnel, some boulders might be broken up during the mining process, and not actually measurable during construction. In such cases the Contractor and Owner may need to jointly agree on a procedure for monitoring boulders encountered.

By establishing clear baselines as a part of the Contract Documents, the parties are more likely to agree on the conditions indicated in the Contract, without time consuming and costly arguments (or litigation) that become counter-productive to a successful project.

The DSC clause provides a mechanism for the Contractor to seek additional compensation due to conditions materially different from those indicated in the Contract. In the question: "Different from what?" the baseline statements define the "what". The more definitive and verifiable the baselines, the easier it should be for the contracting parties to determine the existence of a differing site condition.

As discussed in Chapter 4, Contractors may base their bids on performing the work at a level of difficulty equal to or less adverse than indicated by the baselines. If a Contractor bids below (less adverse than) a baseline for whatever reason, it carries the additional risk associated with that decision. The Contractor has no legitimate basis for a claim if those less adverse conditions are not actually experienced, regardless of whether its assumptions are documented in their bid.

5.2   Contractual Assumptions

Baseline statements in the GBR are assumptions expressed as contractual representations of anticipated geotechnical conditions. A well-written baseline resolves, at least contractually, the uncertainty that may exist in the data or may even extrapolate beyond the range of the data. While the baseline should be realistic and have a rational basis, a reasonable baseline does not have to be based solely upon specific project subsurface information.

The following examples illustrate this point:

- The number of boulders to be encountered may have little to do with how many boulders were identified during the drilling, because the drilling of small diameter holes is not an effective means to detect the presence of boulders. If the designer and Owner consider the risk of encountering boulders to be high, to the extent that it could impact the type of equipment to be used or the manner in which that equipment is outfitted or utilized, the baseline may indicate a greater number of boulders to be encountered than suggested by the borings.

- The potential for slaking behavior of a weak rock may not be based on laboratory test results, but on experiences of nearby projects previously mined within similar geologic formations, and using similar excavation equipment.

- There might be wide scatter in the results of certain rock strength tests; this variability may be related to the quality of the rock samples tested, the manner in which the rock samples were tested; and the availability of a sufficient number of representative rock samples tested. In any case, if it is suspected that the suite of test results might not be representative of the conditions to be encountered, the description of that material's strength in the baseline will probably differ from what could be derived from the data alone.

It is important to provide clear baseline statements. It is also important to describe or present the possible range of property values or material behaviors, for general understanding. The recommended approach is to indicate the expected range of conditions and uncertainty, but then state a specific baseline (upon which bidders may rely) that has been established for contractual purposes. The baseline may be expressed as a maximum value, a minimum value, an average value, a histogram distribution of values, or combinations thereof. The following example illustrates these concepts.

Assume that a tunnel project is to be constructed with a tunnel boring machine through two types of rock; one rock type is stronger and more difficult to bore than the other. The relative percentages of the two rock types along the tunnel alignment are unclear. Given the available information, a reasonable interpretation of the stronger rock to be encountered could range between 30% and 60% of the total tunnel length.

It is almost a certainty that the design team would not correctly predict the actual percentage of stronger rock to be encountered along the tunnel alignment. The recommended approach would be to state the possible range of percentage of stronger rock to be encountered (i.e., 30% to 60%), and then state a realistic percentage to be assumed as the baseline. In this example, that baseline might be set at 45% of the tunnel length. By establishing a clear baseline, the Contractor and Owner both understand the risks to be borne by each; the baseline percentage establishes the amount of stronger rock up to which the Contractor is financially responsible, and beyond which the Owner is financially responsible. The range provides bidders with an informed opinion, so that they may appreciate the level of risk they will take if they base their bid on a set of assumptions less adverse than the baseline (i.e., less than 45% of the tunnel being stronger rock).

If the baseline quantity of stronger rock to be encountered is established at 45%, and the Contractor encounters 50%, and the additional 5% has a quantifiable impact to the extent additional costs were incurred, the Contractor is entitled to additional compensation for the additional 5% strong rock encountered, even though the 50% encountered falls within the range indicated by the data. However, if the baseline is established at 45%, the Contractor bases his bid on 35%, and actually encounters 40% strong rock, there is no basis for a claim. This example underscores the need for a careful mapping and testing of the rock encountered in the tunnel. Effective use of the GBR baseline concept clearly depends on careful documentation of actual conditions encountered in the field during construction.

5.3 Where to Set the Baseline

The baseline can be set, for a given project and set of geotechnical data, at different levels of perceived adversity or difficulty. Where the baseline is set determines the respective levels of risk allocated to the Owner and Contractor. Consider a soft ground tunnel project where it is expected that 100 to 300 boulders could be encountered. An adverse baseline could be set at 300 boulders. The Contractor is obligated to handle 300 boulders and the risk of a differing site condition related to unforeseen boulders is reduced if not eliminated entirely. However, the Owner may pay for the expectation of encountering 300 boulders, whether 300 boulders are encountered or not.

Alternatively, a somewhat less adverse baseline could be set at only 100 boulders. Boulders encountered in excess of the first 100 would be subject to additional payment to the Contractor, through either a pre-determined bid item or a negotiated change order. In this instance, more risk is allocated to the Owner, because additional amounts will be paid if more than 100 boulders are encountered. However, the Owner will probably receive a lower bid, and will only pay an additional sum to the extent that the 100 boulder baseline is exceeded.

Thus, the Owner has an opportunity to exchange higher initial bid prices for a lower number of contract modifications during the work. Regardless of the selected approach, the cost of site conditions remains with the Owner. Risks associated with this issue are discussed further in Chapter 9.

5.4 Baseline Not a "Warranty" of Conditions to be Encountered

The baseline is a representation of what is assumed will be encountered for the purpose of defining "the indications of the Contract". Thus, the provision of a baseline in the Contract is not a warranty that the baseline conditions will, in fact, be encountered. It is therefore not appropriate for the Owner or Contractor to conclude that baseline statements are warranties. However, the baseline statements in a Contract can be considered a contractual commitment by the Owner that those baseline conditions will be applied in the administration of the DSC clause.

This understanding should be addressed in the Contract Documents.

5.5 Link with the Other Contract Documents

Baseline statements in the GBR should be consistent with the design, anticipated construction methods, and measurement and payment provisions in the drawings and specifications. Various means of establishing this consistency are discussed in Chapter 6.

All possible conditions and circumstances that may be encountered cannot and do not need to be included in baseline statements and addressed by measurement and payment provisions. For some conditions, it may be impossible to establish methods of measuring quantities against which payment provisions may be applied. Also, it may prove constructive to require the Contractor to be equipped to accommodate certain potential occurrences, but to treat the payment for such occurrences as DSCs when and if encountered. Examples include the control of groundwater inflows greater than a stated baseline quantity, the encountering, handling and disposal of unknown quantities of contaminated ground and groundwater, or the need for extraordinarily different or additional temporary support of the excavation.

In addition to the need for a close link to other Contract Documents, the GBR offers the opportunity to provide an overview of the project, so that what is contained in the other documents is easier to understand. The drawings and specifications will typically indicate the "what, where, how and when," with little or no justification or explanation. The GBR provides a platform to explain the "why"; i.e., the rationale and bases for items detailed elsewhere. By considering the GBR, all participants to the construction project are provided with an understanding of the key project issues and constraints that have shaped the design and construction requirements. With this background, they are better prepared to understand the rationale behind the requirements of the drawings and specifications, and better prepared to offer innovative ideas for improvements in the form of value engineering change proposals. In some cases, an accepted value engineering change proposal could warrant a modification to the baseline(s) in the GBR.

A careful balance must be sought between providing a document that can be readily absorbed by a bidder without the benefit of having reviewed the other Contract Documents, and paraphrasing the other Contract Documents to the degree of creating ambiguity or contradiction. The GBR should make reference to, rather than repeat or paraphrase, information contained in the GDR, drawings, or specifications.

6.0 PREPARATION OF A GEOTECHNICAL BASELINE REPORT

6.1 Organization and Content

A checklist of items to be considered when preparing a GBR is presented in Table 1. The checklist contains items that permit the GBR to be read as a stand-alone report, without the reader having to refer to discussions or descriptions contained in other Contract Documents. While the checklist in Table 1 is provided within a suggested organizational sequence, other formats may work equally well. The goal is to be clear and concise.

Table 1 covers a broad range and some of the topics will not be applicable to every project. Additionally, the sequence and grouping of topics may be altered to accommodate project requirements or personal preference. For example, a project that has particularly variable conditions across the site and a number of different project components may require the organization and presentation of the anticipated conditions, characterizations, and design and construction considerations separately for each project component. In this manner, the continuity of explaining the key geologic, design, and construction issues for each of the project components may be more effectively maintained.

One important objective in writing a GBR is to produce a concise document that can be read and understood in 2 to 3 hours. For a deep foundation or pipeline project, the GBR will only need to address a few key subsurface and construction issues, and may be as short as 5 to 10 pages in length. A maximum length of 30 pages of text is recommended for straightforward tunneling projects, and no more than 40 to 50 pages for more complicated projects. These page length recommendations can be met while still addressing the items included in Table 1. As explained in Chapter 11, experience in the use of GBRs is that too much is being included in these documents, causing the documents to be so long as to make it difficult to ferret out the baselines. Writers of GBRs are strongly cautioned to avoid overly long or complex descriptions of physical conditions or behaviors. Emphasis should be directed to those physical conditions or behaviors that will most influence the cost of construction or critical equipment to be utilized. Extended geologic descriptions and details should be limited to the Geotechnical Data Report. The writers of GBRs must meet this challenge.

6.2 Writing the GBR – Who, When, and How

The GBR must be prepared by knowledgeable personnel with considerable geotechnical, design, and construction experience relevant to the anticipated project. Owners should retain consultants or consultant teams that include individuals with experience in the local geotechnical conditions, the design and construction of similar projects, and the use of GBRs in the administration of previous construction contracts. Owners should confirm that these individuals will be intimately involved with the preparation and review of the GBR document.

Table 1 - GBR Checklist

Introduction
- project name
- project Owner
- design team (and Design Review Board, if any)
- purpose of report; organization of report
- contractual precedence relative to the GDR and other Contract Documents (refer to the General Conditions)
- project constraints and latitudes

Project Description
- project location
- project type and purpose
- summary of key project features (dimensions, lengths, cross sections, shapes, orientations, support types, lining types, required construction sequences)
- reference to specific specification sections and Drawings to avoid repeating information from other Contract Documents in GBR

Sources of Geologic and Geotechnical Information
- reference to GDR
- designated other available geologic and geotechnical reports
- include the historical precedence for earlier sources of information

Project Geologic Setting
- brief overview of geologic and groundwater setting, origin of deposits, with cross-reference to GDR text, maps, and figures
- brief overview of site exploration and testing programs - avoid unnecessary repetition of GDR text
- surface development and topographic and environmental conditions affecting project layout
- typical surficial exposures and outcrops
- geologic profile along tunnel alignment(s) showing generalized stratigraphy and rock/soil units, and with stick logs to indicate drill hole locations, depths, and orientations

Previous Construction Experience (key points only in GBR if detailed in GDR)
- nearby relevant projects
- relevant features of past projects, with focus on excavation methods, ground behavior, groundwater conditions, and ground support methods
- summary of problems during construction and how they were overcome (with qualifiers as appropriate)
- conditions and circumstances in nearby projects that may be misleading and why

**Table 1 - GBR Checklist
(Continued)**

Ground Characterization
- physical characteristics and occurrences of each distinguishable rock or soil unit, including fill, natural soils, and bedrock; describe degree of weathering / alteration; include near-surface units for foundations/pipelines.
- groundwater conditions; depth to water table; perched water; confined aquifers and hydrostatic pressures; pH; and other key groundwater chemistry details
- soil/rock and groundwater contamination and disposal requirements
- laboratory and field test results presented in histogram (or some other suitable) format, grouped according to each pertinent distinguishable rock or soil unit; reference to tabular summaries contained in the GDR
- ranges and values for baseline purposes; explanations for why the histogram distributions (or other presentations) should be considered representative of the range of properties to be encountered, and if not, why not; rationale for selecting the baseline values and ranges
- blow count data, including correlation factors used to adjust blow counts to Standard Penetration Test (SPT) values, if applicable
- presence of boulders and other obstructions; baselines for number, frequency (i.e., random or concentrated along geologic contacts), size and strength
- bulking/swell factors and soil compaction factors
- baseline descriptions of the depths/thicknesses or various lengths or percentages of each pertinent distinguishable ground type or stratum to be encountered during excavation; properties of each ground type; cross-references to information contained in the drawings or specifications
- values of ground mass permeability, including direct and indirect measurements of permeability values, with reference to tabular summaries contained in the GDR; basis for any potential occurrence of large localized inflows not indicated by ground mass permeability values
- for TBM projects, interpretations of rock mass properties that will be relevant to boreability and cutter wear estimates for each of the distinguishable rock types, including test results that might affect their performance (avoid explicit penetration rate estimates or advance rate estimates)

Design Considerations – Tunnels and Shafts
- description of ground classification system(s) utilized for design purposes, including ground behavior nomenclature
- criteria and methodologies used for the design of ground support and ground stabilization systems, including ground loadings (or reference the drawings/specifications)
- criteria and bases for design of final linings (or reference to drawings/specifications)
- environmental performance considerations such as limitations on settlement and lowering of groundwater levels (or in specifications)

Table 1 – GBR Checklist
(Continued)

- the manner in which different support requirements have been developed for different ground types, and, if required, the protocol to be followed in the field for determination of ground support types for payment; reference to specifications for detailed descriptions ground support methods/sequences
- the rationale for ground performance instrumentation included in the drawings and specifications

Design Considerations - Other Excavations and Foundations
- criteria and methodologies used for the design of excavation support systems, including lateral earth pressure diagrams (or in drawings/specifications) and need to control deflections/deformations
- feasible excavation support systems
- minimum pile tip elevations for deep foundations
- refusal criteria for driven piles
- allowable skin friction for tiebacks
- environmental considerations such as limitations on settlement and lowering of groundwater levels (or in specifications)
- rationale for instrumentation/monitoring shown in the drawings and specifications

Construction Considerations – Tunnels and Shafts
- anticipated ground behavior in response to construction operations within each soil and rock unit
- required sequences of construction (or in drawings/specifications)
- specific anticipated construction difficulties
- rationale for requirements contained in the specifications that either constrain means and methods considered by the Contractor or prescribe specific means and methods (e.g., the required use of an EPB or slurry shield)
- the rationale for baseline estimates of groundwater inflows to be encountered during construction, with baselines for sustained inflows at the heading, flush inflows at the heading, and cumulative sustained groundwater inflows to be pumped at the portal or shaft
- the rationale behind ground improvement techniques and groundwater control methods included in the Contract
- potential sources of delay, such as groundwater inflows, shears and faults, boulders, logs, tiebacks, buried utilities, other manmade obstructions, gases, contaminated soils and groundwater, hot water, and hot rock, etc.

Construction Considerations – Other Excavations and Foundations
- anticipated ground behavior in response to required construction operations within each soil and rock unit

> **Table 1 – GBR Checklist**
> **(Concluded)**
>
> - rippability of rock, till, caliche, or other hard materials, and other excavation considerations including blasting requirements/limitations
> - need for groundwater control and feasible groundwater control methods
> - casing requirements for drilled shafts
> - specific anticipated construction difficulties
> - rationale for requirements contained in the specifications that either constrain means and methods considered by the Contractor or prescribe specific means and methods
> - the rationale for baseline estimates of groundwater inflows to be encountered during construction, with baselines for sustained inflows to be pumped from the excavation
> - the rationale behind ground improvement techniques and groundwater control methods included in the Contract
> - potential sources of delay, such as groundwater inflows, shears and faults, boulders, buried utilities, manmade obstructions, gases, or contaminated soils or groundwater

For smaller projects, where experience with GBRs may not be as well established within the design and construction community, it is particularly important that someone experienced with the preparation of GBRs be retained to either guide its preparation, provide detailed review throughout its development, or both. Some have suggested a form of prequalification to ensure that owners retain professionals with suitable experience in the preparation of GBRs. For example, while having geotechnical engineers involved in GBR preparation is critical, a properly written GBR is substantially different from a more traditional geotechnical or foundations report. Writing an effective GBR represents a challenge for all subsurface projects. Experience suggests that this concern is magnified for smaller projects. Chapter 7 discusses the application of GBRs to other types of subsurface excavations such as deep foundations, open-cut pipelines, or braced or tied-back excavations. It is clear that the success of extending the GBR concept to these types of projects will hinge on the ability to have suitably qualified professionals involved in the process.

An annotated outline of the document should initially be prepared by the geotechnical and design personnel from the design team. This will help focus the report format and content to suit the key project components and construction issues. Sections of the initial draft should then be prepared by the geotechnical consultant, to ensure that interpretations of the exploration results and geotechnical conditions developed earlier in the design process are properly transferred to the GBR. Other sections should be prepared by the design team member who developed the design and prepared the plans and specifications. Close collaboration should be maintained between the geotechnical and design personnel throughout this effort.

All subsequent drafts of the GBR should be advanced jointly by the design team (designer and geotechnical) so that GBR statements are consistent with the developing design, drawings, specifications and payment items. This will facilitate consistency between what is set forth in the GBR, what is contained in the drawings and specifications, and how the Contractor is to be compensated. Advancing drafts should be jointly reviewed by the design team, the Owner, and independent reviewers.

6.3 Link with Risk Registers

Whether under traditional Design-Bid-Build or DB procurement, it is recommended that the GBR be written after most of the design or reference design has been completed. During the site exploration and design phase, Risk Registers should be utilized at earlier stages of project planning, site exploration, and design to help identify key issues. As site exploration, project planning, and detailed design are advanced, certain risks will be identified that are associated with geotechnical and other subsurface conditions. It is precisely those conditions associated with the greatest perceived risks that should be addressed specifically in the GBR. However, as discussed in Chapter 11 there is no need to include the Risk Register in the GBR.

6.4 Wording Suggestions

Baselines are difficult to write without ambiguity. No one can accurately predict the nature and distribution of materials underground and how they will react to excavation. This creates a tendency to use ambiguous words to describe ranges of physical properties and behavior of the materials. The use of words such as "may," "can," "might," "up to," "could," "should," "some", "few", "ranges from ...to...," and "would" are imprecise, and must not be used in baseline statements. Better words include "is," "will," and "are". The use of such definitive terms clearly establishes the intended baselines . As discussed in Section 5.4, the use of these definitive terms must not be taken by the Owner as a warranty by the designer that the underground materials or behaviors are precisely defined. This is the goal of a well-written GBR - to avoid contractual ambiguity.

Whenever possible, baseline statements should be in terms of measurable properties or parameters that can be objectively observed and recorded during construction. The use of adverbs should be avoided. The use of adjectives such as "large," "significant," "local", "many", and "minor" should either be quantified or avoided. If qualitative terms are used, they should be standardized and defined in a summary table or a glossary. As a simple test when writing a baseline statement, ask the question: "If I encountered a site condition pertaining to this baseline would I know if it differed from the indicated conditions?" If a reasonably straightforward affirmative answer is not given, the baseline statement is not sufficiently clear.

Baseline statements regarding anticipated ground behavior should be presented in context with the use of defined means and methods of construction. The baseline statements should make it clear that the ground can (or cannot) be expected to behave differently with the use of alternative tools, methods, sequences, and equipment. In some cases, the Owner may mandate the means and methods and the baselines need to reflect this.

The presentation of baselines regarding groundwater inflows or other phenomena to be measured during performance needs to consider the methods, timing, and responsibilities for measurement in the field. These aspects must be clearly defined and expressed in the Contract Documents.

6.5   Baseline Examples

Examples of problematic and improved baseline statements are presented in Table 2 (the problematic and improved wording are underlined for ease of understanding; baseline words would not normally be underlined in the GBR).

6.6   Consistency and Compatibility

A fundamental shortcoming expressed during the industry forums, is the incompatibility between statements in the GBR and other Contract Document elements and provisions. The GBR should be consistent with and complement the other documents. The following guidelines are useful in achieving these objectives:

- The GBR may present the rationale behind the specification requirements, but should avoid stating the requirements themselves. Detailed requirements should be stated in the specifications only.

- As each baseline statement is prepared and finalized, the technical specifications and payment provisions related to that baseline statement should be reviewed for consistency and reasonableness. For example, if rates of groundwater inflow at the heading are stated as a baseline, the specifications need to define the term "heading", and where and how groundwater inflow measurements are to be taken in the field. If a TBM is involved, these descriptions must consider the physical limitations that will control where a weir or other system for measuring flows may be implemented. Payment provisions included in the Contract for handling and disposing of water must be consistent with the statements in the GBR and specifications.

- The other Contract Documents should be referenced, rather than repeated or paraphrased. If something is stated twice, even only slightly differently, an element of ambiguity is created. As with specifications, the basic rule is "Say it once, and say it well."

- The GBR should explain how the baselines relate to the data contained in the GDR. For example, if the GDR indicates that the maximum unconfined compressive strength (UCS) tested was 19,157 psi, but after discussion with the Owner, the decision was made to require the Contractor to provide for excavating 25,000 psi rock because the strongest rock is seldom found during exploration, an explanation similar

Table 2 - Examples of Baseline Statements

Example	Problematic	Improved
1. Background: A tunnel that is constructed through weak rock that is anticipated to deteriorate when exposed in the tunnel.	The formation is a weak, clay-rich moisture sensitive, soil-like, rock *subject to deterioration upon drying*.	The formation is a weak, clay-rich, moisture sensitive, soil-like rock that *will deteriorate upon drying*.
2. Background: Unstable soils and significant groundwater pressures are expected below the base of a deep shaft excavation.	If groundwater pressures are not *adequately controlled*, the materials in the bottom of the shaft may pipe, heave, or boil, *which could lead to instability of the shaft excavation*.	Unless groundwater pressures are maintained below the bottom of the shaft, the materials in and below the bottom of the shaft *will pipe*, heave, boil, and lead to instability of the shaft excavation.
3. Background: A hard rock tunnel is expected to include three shear zones, each between two and ten feet wide. An open TBM is to be used.	Gripping for TBM thrust reaction *may be somewhat affected* by these conditions, but the effect is *not expected to be severe*.	Gripping for TBM thrust reaction *will be inadequate in these conditions and supplementary thrust reaction must be provided*.
4. Background: In a four-mile long tunnel, two short lengths of bouldery alluvium are expected. Materials less than one foot in diameter are identified as cobbles and are incidental to the excavation. All tunnel excavation and support is bid as a lump sum.	It should be anticipated that *up to* 10 boulders, *as large as* three feet, *could* be encountered in the tunnel.	For baseline purposes, *ten boulders, between one foot and three feet in maximum dimension, are expected to be encountered within soil unit "X"* in the tunnel.
5. Background: Ten shear zones are anticipated to be encountered in a three-mile long tunnel. The definition of "shear zone" is provided in a glossary in the GBR.	*Some* shear zones *may* yield up to 250-gpm initial inflow near the heading, but the flows *should dissipate with time*.	Ten shear zones *are expected to* be encountered in the tunnel. Three of the shear zones are expected to each yield 250-gpm initial inflow to the heading, as measured at the station of the TBM grippers. Regardless of the initial inflow, each of these shear zones *is expected to yield no more than 60 gpm after one month*.
6. Background: A hard rock tunnel is to be excavated using a TBM through massive rock.	The tunnel will encounter granite and granodiorite rocks. The unconfined compressive strength (UCS) of the intact granite *may range from* 6,000 to 25,000 psi; the UCS of the granodiorite *may range from* 8,000 to 35,000 psi.	Granite will be encountered over *40 percent of the length of the tunnel*; the unconfined compressive strength (UCS) of the intact granite will range from 6,000 psi to 25,000 psi, *with an average of 20,000 psi*. Granodiorite will be encountered over the remaining *60 percent of the length of the tunnel*; the UCS of the intact granodiorite will range from 8,000 psi to 35,000 psi, *with an average of 28,000 psi and a range of values as indicated in the histogram in Figure "X"*.

to the following should be provided: "Although the highest UCS tested was 19,157 psi, bidders can anticipate that rock with a UCS of 25,000 psi will be encountered, and that 25% of the total quantity of rock to be excavated will range between 20,000 and 25,000 psi UCS". Stating the baseline in this manner ensures that the project will be equipped to excavate the strongest rock thought likely to be encountered, and establishes as a baseline the quantity of such rock to be encountered. In situations like this, the specifications should also indicate how the strength of such rock is to be evaluated or determined during construction.

- Quantitative baselines should be presented only once. While it is preferable to limit baseline presentations to the GBR, this may not be possible in all instances. For example, if there is a need to indicate the anticipated lengths of different ground types to be encountered in the tunnel, it may be more expedient to show this information on the drawings, with their respective ground stabilization or ground support requirements presented either on the Drawings or in the Specifications. In this instance, the appropriate Drawing(s) or Specification sections should be specifically referenced in the GBR.

- The order of precedence of the different Contract Documents must be clearly indicated in the General Conditions or Special Provisions, to resolve conflicts that may be perceived to exist within the documents. The GBR should take precedence over any other geotechnical report or statement.

While the above may seem axiomatic and reasonably easy to achieve, past performance suggests that the potential for redundancy, ambiguity, and contradiction between the GBR and other Contract Documents is high. Once the drawings and specifications have been finalized, the GBR should be revisited for consistency and to ensure that specification language is not duplicated in the GBR. It is not unusual to revise the GBR four or five times as the drawings and specifications are being finalized. Owners or design managers should not view this iterative process as a negative, but as a critical step toward getting the documents "in sync."

Engaging the "fresh eyes" of independent reviewers in a page-turning process that incorporates the general conditions, technical specifications, drawings, and GBR is the ultimate check on internal compatibility. Because statements in the GBR will be subject to intense scrutiny, interpretation, and possible misinterpretation in the evaluation of potential DSC claims, this independent review of the documents is an essential step in developing an integrated GBR, and is strongly recommended. It is easy to say "make the GBR compatible with the other Contract Documents" but it takes vigilance in practice to achieve it.

6.7 Time and Budget for Preparation

The desired high-quality GBR will not be achieved unless the proper time and budget are allocated to facilitate its development. Preparation of an integrated GBR is as important as the preparation of a set of drawings that is consistent with the specifications. The time and effort involved in making the GBR document internally compatible can rival the efforts expended in preparing the preliminary drafts. An attempt to save money in the preparation of the GBR, either through a truncated or accelerated review process, or the bypassing of a "fresh eyes" review of the Contract Documents, is a false economy - a claim costing the Owner millions of dollars may occur that could otherwise have been avoided.

6.8 Owner Involvement

During preparation of the GBR, meetings should be conducted with the Owner to discuss the topic of baselines. The Owner should be advised of the consequences of an adverse presentation of the anticipated subsurface conditions, versus those of a less adverse presentation, and the need to stay within reasonable limits. The relative implications for how the bid items (or other payment formats) are developed, the initial contract price, potential change orders, and final cost of the work should be carefully reviewed with the Owner, who must be an informed, active participant in the setting of the baselines. The interpretations and baseline statements contained in the GBR should reflect the risk allocation attitudes and preferences of the Owner. The rationale and potential consequences of establishing conservative baselines (i.e. baselines set higher than the data would suggest) should be clearly explained.

7.0 APPLICATIONS FOR OTHER EXCAVATIONS AND FOUNDATIONS

As the application of GBRs has grown in acceptance within the tunneling industry, many Owners have used GBRs for projects or project components involving construction operations other than tunneling. The risks and potential impacts of unforeseen subsurface conditions are just as important for other projects involving subsurface construction. Where construction will involve excavation of the ground that is not visible pre-bid, a set of baseline conditions can reduce uncertainty in the pricing of the work and can serve to assist Contractors and sub-contractors in evaluating the work. Examples of other excavations that would benefit from a GBR include deep building and bridge foundations, open-cut pipelines, braced and tied-back excavations, and highway and other types of earthworks.

As for tunnels, these types of projects are especially prone to adverse schedule and cost impacts as a result of unanticipated subsurface conditions because successful prosecution of the work is focused at one location. Perhaps one of the more ubiquitous applications within public contracting is open-cut pipelines, where disputes, cost over-runs, and construction delays associated with a linear impact zone could be reduced if a GBR were incorporated into the Contract Documents.

7.1 Amplification of Impacts

For the projects addressed here, subsurface risks are often passed-down from the prime Contractor to one or more specialty subcontractors. However, delays to the work can impact the overall project well beyond the specific cost or schedule associated with the subcontract work. Schedule delays, additional overhead, rescheduling, and inefficiencies to the prime Contractor and other subcontractors can result from insufficient information or an inaccurate interpretation of subsurface conditions.

These projects can benefit from the use of a GBR; one that communicates geotechnical risks to the prime Contractor in a manner that can be readily communicated to his applicable subcontractors. The more concise the presentation of the information, the more likely the subcontractors will obtain the information in advance of submitting a bid.

7.2 Baselines for Small Projects

The benefits of the baseline concept are not limited to large projects. Uncertainty regarding subsurface conditions can have a substantial cost and schedule impact on the overall success of smaller projects as well. A delay on an open cut pipeline project can have just as much impact to the public (and the Owner's political circumstances) as a similar delay on a tunnel project.

For projects where the work scope is limited to only a few technical operations, baselines may only need to address one or two key parameters that might affect the equipment requirements, pricing and scheduling of the work. The scope of the GBR should be consistent with the project's size, complexity, and risks. A GBR need not be lengthy in order to be effective.

7.3 Identification of Risk Factors

The key to developing appropriate baselines is to first recognize those aspects of the project that are most dependent on a proper and realistic assessment of the subsurface conditions. As the consequences of geotechnical variations increase, the importance of clear descriptions of key properties and behaviors in the Contract also increases. Critical factors are those that relate to:

- potential construction methods;
- techniques to improve or control the ground or groundwater conditions;
- measurement and payment provisions;
- schedule-related issues;
- environmental and public impacts; and
- design and performance parameters.

The baselines should focus on specific material properties and behavioral characteristics, and should avoid relying on ranges or ambiguous definitions.

Baselines should be relevant to the anticipated construction methods, and readily quantifiable and verifiable using methods that are clear to both the Contractor and the Owner. For example, a recent project described ground conditions as follows:

"The clay in stratum 2 ranges from stiff to hard..."

An improved means of conveying the same information would be:

"As a baseline, the clay in stratum 2 is expected to range in unconfined compressive strength from 2 tsf to 5 tsf, with 25 percent of the clay's strength between 4 tsf and 5 tsf as sampled by ...measured by..."

In the first example the Contractor is left to address several open-ended questions, such as: Is there an upper-bound limit to "hard"? Which published definitions of "stiff" and "hard" should be assumed? How will the ground be sampled and how will strength be measured? In this case, the first example is not a baseline at all, and the Contractor is forced and, arguably, entitled to refer to the data report if included in the contract documents, to develop its own interpretations on soil consistency. In the second example, although the variability of the stratum material is acknowledged, specific strength bounds are established, a baseline that defines the percentage of material within an upper range of strength is clearly portrayed and the method of sampling and measurement is defined. Both the Owner and Contractor can then agree on the details of a field method to assess the soil's unconfined compressive strength throughout the project.

Identification of the appropriate parameters to baseline will be more involved if multiple construction methods are addressed in the project specifications. In this case, the GBR should address each of the anticipated suitable construction methods, and provide baselines (if different) for the various parameters that may affect associated construction under each method, such as the following:

- Stratigraphy
- Material description and classification
- Physical properties (strength, density, moisture content, grain size, plasticity, consolidation, etc.)
- Groundwater elevations and pressures (including perched or artesian)
- Bedrock, till, caliche and other rock-like geologic materials
- Depth of weathering profile and definition of weathering categories
- Buried utilities
- Occurrence of obstructions, both natural and man-made

The list is not comprehensive, and should be evaluated by the design team for each project.

7.4 Baseline Parameters for Consideration

The issues that will impact cost and schedule will vary according to different construction methods. In some cases items to be baselined may deal with construction conditions, obstructions, or hazards. In other instances, the baselines may be represented in terms of design or construction parameters. Table 3 contains a suggested list of items to be considered when developing baseline descriptions for a range of construction methods. The recommendations presented in Table 1 (Chapter 6) apply to all projects, including smaller non-tunneling-related excavations. The purpose of Table 3 is to encourage Owners and designers to consider the application of GBRs for open cut and foundation engineering projects.

Table 3 – Items to Consider When Developing Baselines

Method	Element of Work	Baseline Item
Open Cut Pipelines	Excavation means and methods	• Soil contact elevations • Top of rock or hard material (till, caliche), thickness and characteristics of weathered rock • Rippability of rock with specific methods • Blasting of rock with specific methods and constraints • Use of hoe mounted rock breakers or boom mounted impact hammers • Stand up time
	Groundwater management	• Static water elevation • Location of perched or confined water • Anticipated water inflow rates and volumes • Susceptibility of soils to piping • Water inflow volumes from the backfill around adjacent utilities • Stability of ground below the water table • Need or limits for groundwater control (including feasible groundwater control methods)
	Earth support methods	• Feasible excavation support systems • Limitations – such as limits on driving sheet piles • Lateral earth pressures for the design of excavation support systems; short and long-term
	Excavation methods	• Obstructions and frequency
	Spoil disposal	• Spoil disposal criteria • Amount of material to be disposed by special methods • Influence of Contractor-selected trench width
Excavation and earthwork	Material balance	• Soil stratigraphy, and quantification of material volumes where segregation is required for reuse and/or disposal • Bulking/Swell factors
	Compaction	• In situ soil moisture content • Optimum moisture content • Need for moisture conditioning • Swell/shrink factors
	Rock excavation	• Rippability of rock, till, caliche, etc. • Need for blasting • Bulking/swell factors
	Contaminant disposal	• Extent of contamination and disposal requirements for water and soil
Driven Piles	Installation time Load capacity Equipment selection	• Blow count profiles for use in pile driving analyses • Correlations to be used to convert different sampling methods • Refusal criteria, depth to refusal • Obstructions, including nature, frequency, and distribution

Table 3 – Items to Consider When Developing Baselines (Concluded)

Method	Element of Work	Baseline Item
Drilled Shafts	Casing requirements	• Ground strength • Stand-up time • Water conditions and occurrence of permeable strata • Need for casing or drilling mud
	Groundwater management	• Inflow elevation and pressure • Hydraulic conductivity of specific strata • Inflow volume • Anticipated water or drilling fluid loss rates, disposal issues
	Drilling tool selection	• Material strength and hardness • Presence and number of obstructions • Definition and characteristics of rock, weathered rock, and intermediate degrees of weathering
	Concrete volumes	• Anticipated overbreak
	Soil disposal	• Disposal of muck with contamination
Slurry Walls	Site preparation Obstruction delays	• Obstructions and abandoned utilities – occurrence and frequency • Occurrence and frequency of boulders
	Slurry stability	• Natural pH of the ground, and its effect on slurry
	Desanding systems	• Soil gradation, in particular fines content
	Excavation means and methods Rock excavation rate	• Top of rock, thickness and characteristics of weathered rock • Strength of rock with depth, bedding, or discontinuity orientations • Presence of clay seams, sand pockets, or voids
Ground Freezing	Site preparation drilling of freeze holes	• Obstructions and abandoned utilities – occurrence and frequency • Occurrence and frequency of boulders
	Groundwater hydrology	• Ground mass hydraulic conductivity • Groundwater flow velocity
Tiebacks and Anchors	Material quantities	• Allowable skin friction (w/caution) • Grout take and overbreak (w/caution) • Corrosion protection requirements
	Casing requirements Seal and gasket requirements	• Stand-up time • Water inflow volume and frequency

8.0 DESIGN-BUILD PROCUREMENT

GBRs initially evolved within a traditional Design-Bid-Build procurement framework, where the design was completed by the Owner's consultant prior to a competitive bid process. Under DB procurement, or variations of DB such as public-private partnerships or concessionaire schemes, adjustments to the GBR development process are warranted, but the fundamental concept remains - the Owner owns the ground. The role of a GBR as a risk-sharing tool is equally critical to construction projects under DB procurement as it is under the traditional method.

A significant issue for DB procurement is the means by which the Owner and DB team reach agreement about the geotechnical conditions to be expected during the work. Once that agreed definition of expected conditions is reached, the issue of Differing Site Conditions during construction is handled in the same way as for traditional Design-Bid-Build procurement. This chapter addresses suggested means for reaching that agreement.

8.1 Site Exploration

In traditional contracting, the Owner and his design engineer will address the full scope of geotechnical investigation and design, including exploration of subsurface conditions along the project alignment. Under the DB method, the Owner may seek to transfer the responsibility for portions of this effort to the DB team, whether to achieve schedule efficiencies, transfer subsurface risk, or other reasons.

It is recommended that the same level of exploration be carried out in advance of DB procurement as would be accomplished under the traditional method.
To "economize" on the amount of subsurface information provided in advance of DB proposals increases the risk that the designer will have insufficient information upon which to base a reliable design.

8.2 Geotechnical Data Report

The philosophy with regard to geotechnical data reports is substantially the same for DB procurement as for traditional procurement. It is incumbent upon the Owner to assemble all data and information that has been obtained in the course of the site characterization effort, and to disclose this information in an organized fashion. The information should be organized and presented in a meaningful format in a Geotechnical Data Report (GDR) in much the same manner as for traditional procurement methods. It is imperative that all such factual information be incorporated into the Contract Documents, so that the DB team has an appropriate database upon which to rely in the development of their design and in selecting their means, methods, and excavation approaches.

Bidders in a DB procurement process should be afforded the opportunity to obtain additional information at locations critical to their planning and design. Providing a background understanding of exploration gaps and constraints will help guide bidders in understanding what might be accomplished through supplementary exploration requests.

The results of any exploration and testing carried out by the Owner during the bid process should be included in the GDR, and made available to all the DB teams.

8.3 Geotechnical Baseline Report

Under DB, although the Owner supervises the gathering of the subsurface information, design-specific interpretations and decision-making lie with the DB team. The content of GBRs for DB projects should be substantially the same as described in Chapter 6; however a modified process that allows the DB team to participate in the development of the GBR is required.

Several projects have incorporated a GBR into a DB contract, as discussed in Section 8.4. Drawing on these and other experiences, the following three-step approach is suggested:

Step 1 – GBR-B. On the basis of the site exploration program and preliminary design, the Owner (through its geotechnical and design team) prepares a Geotechnical Baseline Report for Bidding (GBR-B). The focus of this document is the *physical* nature of the subsurface conditions likely to be encountered, consistent with the layouts and geometries represented in the preliminary design. In this manner, all Design-Build teams are provided with the same set of physical baseline conditions to be used in their design and construction planning. The document should:

- describe the bases for the preliminary designs provided by the Owner's design team;

- provide key baselines of anticipated physical conditions consistent with the exploration program and other relevant construction; and

- to the extent desired by the Owner or required by third-party constraints, mandate or preclude the use of certain equipment, means, and methods.

The degree to which the GBR-B provides *behavioral* baselines will be a function of the level of specificity in the preliminary design. It would be inappropriate for the Owner's design team to address behavioral issues in detail because such issues will be closely linked to the equipment, means and methods selected by each DB team. Different construction approaches may warrant different geotechnical considerations, and therefore warrant different behavioral baselines. Some examples are provided for illustration.

In a hard rock tunnel, viable alternatives might include drill and blast methods, or the use of a roadheader or tunnel boring machine (TBM). Different rock mass characteristics will influence the efficiency of the three different excavation methods, as well as the behavior of the resulting excavated openings and required ground support. Certain rock defects, such as bedding planes, joints, or shears, may have a relatively minor impact on the advance rates achieved by drill and blast methods but will have a more significant impact on overbreak, the need for initial support, and the cost of providing a final lining. In contrast, those same rock mass defects, depending on their orientations, may have more influence on the advance rate (penetration rate) of a roadheader or TBM, but have less influence on overbreak during excavation or the extent of initial support required. Thus, a discussion of rock mass parameters and behavioral characteristics must be described within the context of the various anticipated methods of tunnel excavation.

In a soft ground tunnel, the effectiveness of a cutter wheel machine might depend on different ground characteristics than those that would influence the effectiveness of an open-face digger shield. Boulders might represent a problem for the cutter wheel machine but not for the open face shield. Zones or pockets of unstable ground might influence each excavation method to a differing degree. If the use of a pressurized face machine is mandated, soil types, grain size distribution, permeability, and groundwater conditions will most likely have a different impact on an earth pressure balance machine as compared to a slurry pressure machine. Again, the specific soil characteristics and behaviors important to project success will vary with the specific means and methods.

In order to facilitate the comparison of documents from multiple teams, a common format is achieved by having the GBR-B prepared with discrete sections of the report left blank. The blanks contain annotations prompting bidders to address these specific issues and behavioral aspects consistent with their chosen equipment, means and methods.

The Owner should update the GBR-B during the bid process to reflect the results of any additional exploration and testing carried out by the Owner during that time.

Step 2 – GBR-C. As a part of their detailed design and construction planning process, each DB team will interpret the various baselines expressed in the GBR-B, consider those baselines in the development of their design and construction approaches, and fill in the gaps and blanks in the GBR-B accordingly. Consideration could be given to distributing the GBR-B in electronic form, so that any modifications or clarifications suggested by each DB team are captured in the track-change mode of most computerized word processing software programs. In its completed form, the

GBR for Construction (GBR-C) will reflect the physical baselines established by the Owner and its design team (as augmented by any supplemental exploration) and as clarified or modified by the DB team, and the behavioral baselines described by the DB team consistent with it's design approach, equipment, means, and methods.

Step 3 - Owner Review and Negotiation. As a part of the negotiation process, the Owner should have the opportunity to review each DB team's GBR-C for concurrence and reasonableness. If, in the Owner's (or its design team's) opinion, the baseline assumptions prepared by a particular team are judged to be optimistic, vague, or otherwise incompatible with statements in the GBR-B, the Owner should seek clarifications through discussion with that team. If those clarifications have an influence on the cost of the work, the DB team should be given the opportunity to revise their pricing, adjust the payment terms or provisions, or a combination thereof. The Owner may also choose to carry out such negotiations with more than one DB team. After the Owner and the successful DB team agree on such changes, the modified GBR-C supersedes the GBR-B and is incorporated into the DB contract. From that point forward, its use and function is similar to that for a GBR within the traditional contractual framework.

8.4 Recent Applications

A number of DB projects have been implemented in recent years with some variation of the above-described approach. **The Tren Urbano Subway** in San Juan, Puerto Rico provided a database of geotechnical information, and required each bidder to propose supplemental investigations to be carried out by the Owner during the bid period. No interpretations were submitted within the context of baseline statements or descriptions. Supplemental information gathered by the Owner during the bidding process was shared with all bidders prior to the submittal of final bids. Each bidder was required to prepare a Geotechnical Design Summary Report (GDSR) and to submit their GDSR with their bid. The Owner utilized this information during the bid review to gain an understanding of each team's perceptions of the risks on the project. Portions of the winning team's proposal were excerpted and included in the DB contract. Their GDSR was retained as a part of their Bid Escrow Documentation. The Contract included a Differing Site Conditions clause and a Dispute Review Board.

For the **Deep Tunnel Sewerage Scheme** in Singapore, a different approach was taken. The Owner's engineering consultant provided a Geotechnical Data Report only, with only modest interpretations of the anticipated subsurface conditions for each of a number of different construction contracts. As a part of the tender process, each DB team was required to prepare a Geotechnical Interpretive Report (GIR), which was required to address specific issues, estimated behaviors, and anticipated parameters influencing tunnel heading advance rate. The Owner and their consultant critically reviewed the interpretive reports during the evaluation of tenders. The contract included a form of a Differing Site Conditions clause, and the selected team's

GIR was incorporated into the DB contract. The number of borings drilled in advance of the tenders was less than suggested by industry standards, and each team was required to price a certain number of supplementary borings at specific locations. Unfortunately, because timetables for the selection of means and methods preceded the completion of the supplementary borings, certain equipment and design decisions were made without the benefit of the additional information.

Seattle's Sound Transit Program pursued yet another approach to ground characterization for a DB project that was bid but never awarded. The Owner's approach built upon experience gained during the Tren Urbano project, and engaged the tunneling community in seeking new ideas. The north corridor portion of the project, which was to include about 5 miles of twin-bore tunnel, was to have been constructed following a DB format. A phased exploration program was completed during the feasibility and preliminary design phases of the project. Borings were spaced an average of 330 feet apart along the tunnel alignment. In addition, six to nine borings were drilled at each of four transit station sites. All exploration and laboratory data, as well as a discussion of local tunnel case histories were presented in a GDR and interpreted in a Geotechnical Characterization Report (GCR).

Three DB teams were pre-qualified, but one withdrew during the proposal phase. An additional four borings were drilled at the suggestion of the two remaining teams. After the selection of the preferred team, an additional 15 borings were completed based on discussions with that team. All additional exploration information was presented in an appendix to the GDR. The Owner's geotechnical engineer prepared a Tender Geotechnical Baseline Report (TGBR) that established baseline conditions for relevant physical conditions such as boulder quantities, the nature of the soil units to be encountered, and baseline behavior of the soils referenced to an unsupported face condition. All three reports were to be included in the DB contract (Robinson, et al., 2001). The teams were requested to factor the contents of the TGBR into their bids, and to write a companion document to the Owner's TGBR that documented their selected means, methods, and associated ground behavior assessments. The intent of the process, had it been fully completed, was for the Owner and preferred DB team to reach a consensus understanding of a joint GBR, which would have been incorporated into the DB contract. The process did not advance to the final stages of negotiation due to funding issues.

The Niagara Tunnel Project in Ontario is a DB project that closely followed the three-step process described above. A tender document entitled a GBR-A was prepared by the Owner and submitted as a tender document to prequalified DB teams. The GBR-A contained gaps in the form of comment boxes that solicited input from each team. The document included with each team's proposal was referred to as their GBR-B. Though the construction approaches and designs were different among the different teams, the Owner was able to compare the different GBR-B assumptions and assess compatibility with different design and construction approaches. The selected team's GBR-B was discussed, modified, and agreed upon by the parties prior to incorporating it into the Construction Contract as the GBR-C. The Niagara Tunnel Project, which is currently under construction, includes a Disputes Review Board and a form of a Differing Site Conditions clause.

The San Diego County Water Authority utilized a DB approach for the **Lake Hodges to Olivenhain Pipeline - Tunnel and Shaft**. This project involves the construction of a 5,800 foot-long tunnel and 200 to 800 foot-deep drop shafts. A GBR-B was prepared by the Owner's engineer and issued with the bid documents. After Notice to Proceed, as part of their design development, the Contractor was required to prepare a GBR-C incorporating interpretations of any new geotechnical data obtained and providing assessments of ground behavior and ground response for the selected means and methods. The project is currently under construction and this approach to the development of the GBR appears to have been effective.

The **Sea to Sky Highway Improvement Project** is a DB roadway project that utilized a baseline approach on one portion of the project alignment. The Public-Private Partnership undertaking involves the widening and upgrading of 39 miles of two, three, and four-lane highway between Vancouver and Whistler, British Columbia. No tunnel construction is involved. The work involves drilling and blasting of cut slopes, the construction of downslope retaining walls and viaduct-like structures, and multiple bridge crossings. The Owner provided all pre-bid subsurface information to the bidders in a data room. For the majority of the project, the DB team is responsible for all geotechnical and subsurface conditions.

However, for a particularly challenging 7-mile portion of the alignment, the DB team developed and presented a set of design assumptions relating to the presence of rock, and the presence and depth of rock fill. These assumptions were recognized by the Owner and DB team as contractual baselines. During the execution of the work, if the subsurface conditions are found to be more onerous than baselined, the DB team and Owner have a means of negotiating design changes to mitigate additional construction costs. The residual impacts are shared equally between the Owner and DB team. The DB team retains the financial benefit of more favorable conditions.

A more collaborative GBR development process can be realized under DB procurement. It is considered that the more collaborative the process, the more effective the resulting product will be in helping to avoid and resolve disputes. The key is for both parties to fully understand and agree what the baselines mean, and how changes will be addressed if different conditions are encountered during the work.

9.0 OWNER PERSPECTIVES

9.1 Realities in the Public Sector

The Owner of an underground project must deal with certain issues that the Contractor and designer may not readily appreciate. One common issue is that funds are generally limited. A public project must compete with others, many times in a political venue, for the available capital funds. When the competition and demand for funds is high, the "budget" for a given project may be redefined as the project moves through preliminary design, final design, and construction bidding. When the design of the project is being developed, the "budget" is the sum of the designer's estimate and contingencies. However, once the project bids, the "budget" often becomes the amount of the awarded contract; previously included contingency funds may be reassigned to other public works projects.

Some Owners may be able to maintain a percentage of the contract award amount as contingency funds. As high as 10 to 50 percent of the contract amount, depending on the design and geotechnical risks anticipated, may be set aside. Many Owners, however, are not that fortunate.

Requests for additional funds often present a degree of political risk for the Owner's project manager or project manager's supervisor, to the extent that there is resistance to seek additional funds during construction. In certain organizations, the performance of the Owner's project manager may be judged on his ability to avoid "cost overruns." Thus, the Owner may prefer to have a baseline that attempts to minimize project change orders even at the cost of higher initial contract prices. Alternatively, the Owner may elect to include specified allowances or provisional funds in the initial bid price to be utilized, if required, for contract adjustment.

The Owner may also prefer to specify less risky but more costly designs and construction procedures to avoid politically undesirable events, such as the risk of excessive settlements of public streets or adjacent buildings. In this event, the rationale for these requirements should be clearly explained so that the bidders will be able to understand the reason for the "conservative" approach.

9.2 Setting the Baseline

Chapter 5 describes the concept of the baseline, and explains that different baselines can be developed considering the same geotechnical data. Where the baseline is set determines risk allocation and has a great influence on risk acceptance, bid prices, quantity of change orders, and the final cost of the project.

Owners should participate in and contribute to the setting of the baselines and should understand the consequences of the levels at which the baselines are set. The design team should explain the possible baseline range (with a discussion of the associated risks) and offer a "most reasonable" interpretation for the Owner's consideration.

A baseline that portrays a relatively adverse site condition will tend to:

- increase the bid price by allocating more risk to the Contractor;
- allocate less risk to the Owner and reduce the potential for change orders; and
- cost the Owner more, due to paying for the contingency of encountering the adverse condition, whether or not the condition is actually encountered.

A baseline that portrays a less adverse site condition will tend to:

- decrease the bid price by allocating less risk to the Contractor;
- allocate more risk to the Owner and increase the potential for change orders;
- cost the Owner more than the bid price if the adverse site condition(s) is encountered; but
- cost the Owner less if the adverse site condition is not encountered.

If the condition actually encountered requires a significant modification of the means and methods, the associated cost and schedule impacts will likely be greater than establishing a more adverse baseline in the first place.

In cases, a) a relatively adverse baseline, and b) a less adverse baseline, the cost of dealing with subsurface conditions ultimately rests with the Owner. Varying the baseline does not shed that responsibility. The difference between the two cases is whether the Owner a) pays a premium at the beginning of the Contract with higher bid prices (irrespective of whether the adverse condition is actually encountered); or b) pays only if and when it is actually encountered during construction, either by change order or some other provision under the Contract.

The attempt to eliminate claims through an overly adverse baseline may have the opposite effect if the baseline is unrealistic. Bidders will recognize an unrealistic baseline, and in order to be competitive, will be inclined to base their bid upon a more realistic condition, in effect, below the baseline. Then, if the more adverse condition is actually encountered during construction, a claim may still be filed with the argument that the "baseline" was unrealistic. As discussed further in Chapter 11 the risks associated with bidding below the baseline lie with the Contractor, however the Owner may suffer through an expensive adjudication process to prove the point. The possible benefits of arbitrarily adjusting the baseline for selective allocation of risk must be given careful consideration. Usually the best approach is to set a realistic baseline and incorporate contingencies for adverse conditions. Deviation from this principle will often lead to problems of "unintended consequences".

Baselines that strictly reflect the data and interpretations developed during site exploration and design may not be the best choice from the Owner's perspective if other information suggests that the database is incomplete or non-representative. The design team should convey to the Owner the potential consequences of setting the baseline at different levels of adversity, as they relate to:

- the effect on the bid price;

- the potential for change orders relating to differing site conditions; and

- the likely overall cost of construction.

The Owner needs to appreciate the inter-relationships between the above factors, the limitations involved with the characterization of ground conditions and ground behavior, and participate in the discussions and considerations that precede the setting of the baselines. Only then will baselines, and the resulting allocations of risk, reflect the Owner's desires relative to management of change orders, requests for additional budgets, and management of overall construction costs.

When considering the Owner in the process of setting baselines, it is imperative to differentiate and anticipate who the "Owner" is during the design phase as opposed to the construction phase. During the design process, when baseline decisions are made, the Owner's interests may be represented by individuals from the design or engineering branch. However, when the consequences of the baseline decisions are revealed during construction, the Owner's interests may be represented by more senior officials, board members, or the legal division who have little or no relevant construction or engineering experience, or construction managers who had no involvement during the design. The Owner's interests will be best served if these representatives either participate in setting the baselines, or are informed of earlier baseline decisions before the contract goes out for bid.

9.3 Managing the Owner's Risk

The Owner is understandably concerned about managing the financial risk throughout the construction process. Four realities that the Owner must understand and appreciate at the outset, through careful advising by the design team, are:

- Accurately anticipating every ground condition is rarely possible. Variations and unexpected conditions will occur.

- Construction risk should be allocated fairly; Owners must deal with the risk of unanticipated subsurface conditions. This risk cannot be eliminated.

- A baseline is no guarantee against differing site condition claims, or against the need to adjust pay quantities for unit price work during contract performance.

- Exceeding the baseline, in and of itself, does not represent a defective design.

Owners should understand what they can do to reduce their risk. One measure is to provide adequate schedule and budget to explore the subsurface conditions, not only for the designers' purposes, but also for bid preparation and construction purposes. There is no substitute for carrying out a comprehensive site exploration program. The more that is known about the ground conditions, the more accurate the anticipated cost of the project will be. If there is an area of identified risk that can be better managed or understood by seeking additional information in a supplementary exploration program, Owners should be willing to invest the time and money to carry out such additional investigations. This is money well spent.

A second measure is to retain suitably qualified and experienced geotechnical and design consultants with prior local geotechnical and construction experience to investigate the subsurface conditions, to evaluate the potential risks, and to prepare internally consistent drawings, specifications, and GBR.

A third measure is to allocate sufficient budget and time to permit proper preparation and review of all of the contract documents, including the drawings, specifications, and GBR. If an independent review board is utilized, it is beneficial to give the board the opportunity to review and provide comments on at least two drafts of the GBR together with the drawings and specifications. This will better ensure that the documents are internally consistent. When the exploration and design process are accelerated to meet predetermined deadlines, the result may be a sub-standard GBR, which could increase the final cost and schedule of the project. An additional month or two at the end of a multi-year design schedule will mean little to the overall schedule but may have a profound impact on the quality of the final Contract Documents. If a separate Construction Manager (CM) is to be used for the project, the Owner should engage the CM during the design process, not only to provide constructability review of the design, plans, and specifications, but also to participate in review of the GBR.

A fourth measure is to develop unit price payment provisions that can accommodate the full range of anticipated conditions and reasonable variations in ground conditions. By including these items in the bid schedule, competitive prices for these items are obtained during the bid process. Bid items for different degrees or quantities of items, such as groundwater inflows, ground support, grouting, etc., provide an effective means of dealing with those conditions if and when they are encountered, and minimize or eliminate DSC claims. A bid item for delay time, with a fair preset price or high enough quantity to preclude unbalancing, provides a means to deal with completely unanticipated conditions.

A fifth measure is to minimize misunderstandings as to what is indicated by the GBR, by discussing the baselines and encouraging candid discussion of the GBR contents with the bidders at the prebid meeting. This may be more easily achieved with private Owners than in the public domain. However, the ability to eliminate uncertainties among the potential bidders prior to the submission of bids will resolve many questions that might otherwise lead to unanticipated change orders and disputes.

The Owner can manage exposure to additional construction costs by maintaining a contingency fund apart from the construction contract. This fund should be maintained until all the potential design and geotechnical risks have been adequately addressed. An appropriate contingency fund should reflect the complexity of the anticipated conditions and the Owner's perceived risk. This is usually much higher for tunnel projects than for other types of construction.

10.0 ROLES AND RESPONSIBILITIES

Though the various roles and responsibilities of the parties have been discussed previously, it is useful to summarize them here.

The Owner should:

- provide adequate funding and schedule for geotechnical exploration and for preparation and review of the GBR;

- participate in the process of setting baselines, both to understand the risks and risk allocation and to fully understand and approve the baseline statements;

- thoroughly review and understand the various iterations during GBR development, including the review of DB GBR-C submittals;

- understand the vagaries of subsurface construction, and maintain an adequate reserve fund until all potential risks have been adequately addressed;

- provide sufficient budget during construction for adequate documentation of the actual conditions, so that the parties can agree on the conditions that were encountered, and the circumstances under which they were encountered; and

- promptly compensate the Contractor for valid DSC claims.

Under Design-Bid-Build procurement, the design team (geotechnical engineer or engineering geologist, and design engineer) must:

- provide geotechnical engineers and engineering geologists experienced in site investigations, data collection, and report preparation for the type of construction project being undertaken;

- prepare interpretations of the data that address design and construction concerns for geotechnically feasible design options;

- provide geotechnical and design engineers experienced in the appropriate type of design and construction to prepare and review the plans, specifications, and GBR;

- inform and educate the Owner as to the purpose and use of baselines;

- write clear, concise, definitive, and realistic baselines that are compatible with the drawings and specifications;

- explain baselines that are different than indicated by the data;
- explain the baseline statements and their consequences to the Owner and Contractor;
- write baselines that can be objectively evaluated; and
- indicate how baseline conditions will be measured and evaluated in the field.

Under DB procurement, the Owner's design team must also:

- provide a thorough review of the various GBR-C documents from the DB teams;
- explain to the Owner any differences that may exist between the different proposals; and
- assist the Owner in negotiating agreeable wording in what will become the standing GBR to the DB contract.

The Contractor must:

- seek clarification of unclear contractual provisions before bid;
- bid the work with a clear understanding of the GBR information, the contractual baselines, and his interpretation of the anticipated geotechnical conditions;
- bid the work with full consideration of the available geologic data in the GDR information if the bid is based on innovative or unusual equipment, means or methods, or if the bid is below the baseline(s);
- share the GBR and GDR information and interpretation of ground conditions with its major equipment suppliers, subcontractors, design consultants, and materials suppliers;
- understand and accept the level of risk associated with his bid assumptions that are less adverse than the baselines;
- accept the responsibility for selection of means and methods of construction and their impact on ground performance;
- provide means, methods, and equipment consistent with the baseline conditions and other indications in the Contract; and

- promptly make required adjustments if the initially selected means and methods are inappropriate.

Under DB procurement, the Contractor must also:

- retain its own design team to assist with preparation of the GBR-C and if required, help negotiate and finalize the standing GBR.

The Construction Manager should:

- be given the opportunity to participate in the review of the GBR during its preparation;

- fully document the actual conditions encountered (particularly those described in the baselines) and the impacts of such conditions on the construction;

- carefully and thoroughly evaluate DSC claims submitted by the Contractor;

- acknowledge the existence of, and encourage the Owner to promptly compensate the Contractor for, valid DSCs; and

- when appropriate, firmly and convincingly explain to the Contractor why a particular DSC claim is not valid.

Finally, if called upon, the Dispute Adjudicators must:

- make interpretations using the Contract as a whole, and in the event of conflict respect the contractual hierarchy of the Contract Documents;

- apply the baselines as stated in the GBR, e.g., refrain from invoking judgments that conflict with stated baselines;

- take into account the influence of the Contractor's means and methods, workmanship, and efficiency on ground behavior and overall performance/progress;

- recommend entitlement for conditions more adverse than the baselines only if they have resulted in material additional costs to the Contractor; and

- deny the merit of claims if encountered conditions are shown to be consistent with or less adverse than the conditions described in the GBR.

11.0 LESSONS LEARNED

The first edition of this document was published in early 1997. In the following ten years, many projects have been planned, designed and constructed. Likewise, many disputes have been argued and reconciled, and a multitude of lessons have hopefully been learned. One way to improve the business practice is to capture these lessons learned and share them among practitioners.

This chapter discusses feedback obtained during two industry workshops that were specifically directed at capturing lessons learned. The first workshop, held in association with a national tunneling conference in 2004, consisted of an all-day program dedicated to the review and discussion of items regarding GBRs that a six-person panel considered the most controversial. The second workshop, held in association with a national tunneling conference in 2006, involved industry review and commentary on a draft manuscript of this publication.

11.1 GBR Preparation

The general observation from both workshops is that the GBR concept is working well and that properly expressed baselines and GBR documents will go a long way to enhancing the effectiveness of GBRs. Authors need to do a better job of anticipating construction conditions, and then explaining those conditions in clear, concise, and unambiguous terms. It has also become increasingly evident that a poorly written GBR, or a GBR that is poorly integrated into the other Contract Documents, can become a lightening rod for claims and disputes.

As a tool for avoidance and minimization of disputes related to subsurface conditions, the GBR is the single most important document in the Contract. It is typically one of the first documents a bidder will review, and likely the first that an adjudicator will assess when presented with a dispute.

The importance of GBRs being prepared by suitably qualified professionals, and for the GBR and other Contract Documents to be reviewed by the "fresh eyes" of professionals experienced in the types of construction under consideration cannot be over emphasized. It is not unusual to have the GBR go through eight to ten reviews throughout the design and development of the Contract Documents. This is not an indication of having it wrong – it is a requisite for getting it right.

11.2 April 2004 Workshop

The Workshop held in Atlanta in 2004 consisted of discussions among a panel of representatives including Contractors, Owners, engineering consultants and attorneys, along with feedback and commentary by the audience. Each panelist addressed what they considered to be their three top issues and concerns regarding the preparation and interpretation of GBRs within a construction contract. The program Chairman led the panelists through a point-counterpoint discussion of each issue. The audience participants were encouraged to voice their views regarding the panelists' issues, as well as to raise any additional observations or concerns.

A broad array of issues was addressed, many focusing on "nitty-gritty" topics that have arisen in the course of dispute resolution on multiple projects. The more significant issues discussed at the workshop are summarized below.

Are baselines only enforceable if based solely on the available data?

In most instances the available exploration database is not sufficiently complete to fully characterize the anticipated subsurface conditions. In this instance, the GBR must go beyond the available data to provide a reasonable baseline of anticipated conditions. Examples include: boulders in the baseline when no boulders are actually detected in the soil boring program but generally known to exist; complex depositional environments not sufficiently documented in the borings, but known to exist based on the geology and/or experience from previous construction contracts in the area; the existence of vertical shears or faults that may have been missed in the boring program for a variety of statistical and geometric reasons; and adverse rock mass or groundwater conditions known or thought to occur at formation contacts, but not well documented in the boring logs.

The consensus is that such interpretations and extrapolations from the available geotechnical data should be reasonable, and should be explained if the resulting baselines deviate from the available data. Such modifications should be thoroughly vetted and resolved during the pre-bid and pre-award period, and not left to form the basis of potential disputes during actual prosecution of the work.

Baselines are contractually binding, regardless of the presence or absence of specific substantiating geotechnical data. The "baseline" is the "baseline".

Should dispute adjudicators respect the baselines when hearing disputes?

The majority of contracts that include GBRs also include a Disputes Resolution Board (DRB) as part of the disputes resolution process. As stated above, the baselines are binding on both parties and must be respected by the DRB.

Early DRB guidelines used the terms "fair" and "equitable" in reference to resolution of disputes. Some DRB members may have misunderstood these terms to mean that the DRB had the authority, if not the responsibility, to correct seemingly "unfair" or "inequitable" conditions contained in the Contract Documents. However, by bidding the work, the terms of the Contract, including the baselines, were acceptable to all bidders.

Recent DRB guidelines do not contain the terms "fair" and "equitable". Rather, the guidelines state that disputes brought to the Board are to be resolved "on the basis of the facts in the case, the terms of the contract and prevailing law" and, further, that the DRB does not have the authority to alter the terms of the Contract.

The lessons learned here are that: (1) GBRs should be written by knowledgeable professionals with clear and precise baselines that minimize the potential for misinterpretation; (2) Contractors should price the work with full consideration of the risks allocated by the baselines; (3) DRBs and other dispute adjudicators should interpret the GBR as a singular document, not as multiple sections with different priority levels; and (4) DRBs and other dispute adjudicators should apply the baselines in the GBR in accordance with the Contract Documents.

Should baselines be precise values, or can/should they be ranges?

If the variability of a material property or characteristic is legitimately reflected in the range of data, it is considered appropriate to state: "The available information indicates that item Q will range between X and Z; and for baseline purposes assume that the average Q = Y". This satisfies the desire and appropriateness of communicating the uncertainty, while providing a clear contractual baseline. As addressed in Chapter 6, a histogram presentation of the available data helps to clarify the anticipated variation in material property or characteristic from the baseline average.

"Soft" baselines (stated as a range of material properties or characteristics, rather than an average) run counter to the general proposition that baselines seek to enhance clarity and reduce ambiguity. A baseline "range" serves neither of the above objectives. Before the development of baselines, only numerical geotechnical data were provided, and usually in the form of ranges. The concept of stating baselines as a range is a step backward, not forward. Proponents of baseline ranges defend it by maintaining that the range represents an "uncertainty zone." While that might be true, the job of the GBR is to wring the geotechnical ambiguity out of the bidding process, not add to it. Excessively wide range-expressions are inconsistent with the overall goal of minimizing disputes. In reality, the use of ranges in a GBR results in the creation of an unnecessarily wide "battle zone".

Some Owners and GBR authors have expressed a concern about being "wrong" if an average property or characteristic is presented as a baseline. If an Owner or Owner's representative is concerned that presenting an average condition as a baseline will lead to a number of minor claims, they have the latitude to adjust the baseline to represent a more adverse condition, recognizing that in exchange for the more conservative baseline, they should anticipate higher bids. Adjusting the baseline to a more conservative value is viewed as a better solution than presenting the baseline as a wide numerical range.

This said, Owners and Engineers are strongly advised to present reasonable, realistic baselines. Building in excessive conservatism not only frustrates the purpose and role of the GBR but will inevitably cause confusion and in the end, may cause the very claims and lawsuits they were trying to prevent.

Can a bidder interpret the Contract payment quantities as a baseline?

Quantities in a bid schedule are frequently interpreted as indications of anticipated geotechnical conditions.

From a legal perspective, under the Federal Differing Site Conditions clause discussed in Section 4.2, a claim for relief due to adverse unexpected site conditions depends on the claimant being able to show that the conditions differ materially from those "indicated in this contract." U. S. construction case law generally supports the interpretation that planned contract payment quantities in the bid schedule are valid "indications" of physical conditions to be expected during construction.

Therefore, while payment quantities are similar to baselines in that they are "indications" of conditions of the Contract, they are not the same. Payment quantities represent the amount of work to be performed by the Contractor and baselines are a description of the physical conditions involved with the work. Nevertheless, payment quantity provisions that are related to certain physical conditions should be compatible with baseline statements made in the GBR. This is precisely the clarity and consistency that is sought in the proper preparation of a GBR. Compatibility with payment provision statements and quantities will serve to eliminate ambiguity, confusion, and disagreement.

Are unreasonably onerous or conservative baselines binding, or should DRBs and other adjudicators ignore them?

As discussed above, the baselines are a part of the Contract, they are binding, and they are not to be ignored or otherwise set aside.

How large a variance is required from the baseline to justify additional compensation?

The answer to this question depends on the particular circumstances - how the baselines are stated, the nature of the encountered conditions, whether the conditions can be deemed a "material difference", and whether an adverse financial impact can be clearly demonstrated.

The concept of what constitutes a "material difference" in the contract definition of a differing site condition, whether physical or behavioral, is central to the dispute resolution process. If there is no demonstration of material difference, there can be no compensation in either time or money.

Should the GBR be presented as a series of discrete statements?

This perspective is offered by those who are concerned that GBRs are becoming too long, and that as the length of a GBR increases, the clarity of what is really important is diminished. However, to reduce the GBR to a list of short, concise statements is equally ineffective. Doing so would exacerbate other expressed shortcomings in GBRs. For example, as discussed earlier in this chapter, it is important to provide the context of a numerical baseline; it helps bidders understand the risks they are accepting if they bid below the baseline. Also, it is important to recognize that the GBR is the sole interpretive geotechnical report provided to the Contractor that has contractual significance. In this context it is fundamental that the GBR help the Contractor understand the project, its challenges, and critical issues, in particular where the geologic conditions are involved. A brief description of limitations of the site characterization effort, clarified with specific baselines can be very effective in communicating the project realities to the Contractor. Removing this background information serves to undercut a full contextual understanding of the conditions and impairs the bidders' ability to assess the level of risk they are being asked to accept, and it is recommended that this not be done.

In summary, the longer and more complicated a GBR, the more likely it is that its message will be diluted and the greater the chances for internal inconsistency. For this reason, authors of GBRs are advised to keep the GBR concise and to the point.

11.3   June 2006 Workshop

A draft of this document was circulated to interested parties, and was discussed in an open forum during a workshop held in conjunction with a national tunneling conference in Chicago in 2006. The five-hour workshop was audio recorded, and a transcript was prepared to capture individuals' comments.

Discussions during this workshop reinforced a number of items raised during the April 2004 Workshop. Positive feedback on two projects ratified the three-step GBR process for DB applications as described in Chapter 8, specifically the Lake Hodges project discussed in Chapter 8, and the New York Avenue Extension of the Washington Metropolitan Area Transit Authority in Washington, DC.

A number of suggestions raised during the workshop were subsequently incorporated into the text of this document. Comments regarding additional topics are provided below.

The Use of Risk Registers

The use of Risk Registers as a formalized risk assessment tool has become more common in recent years. A question was raised about possible incorporation of discussion of Risk Register evaluations into the GBR.

By its nature, the Risk Register process addresses a much broader range of risks than those addressed in a GBR, and discusses specific mitigation strategies. For an underground project, the outgrowth of such mitigation strategies may have already been incorporated in the design, project alignment, selection of construction methods, etc. Thus, the goals of the Risk Register process may have already been achieved. Nevertheless, there is a logical connection between the identification of geotechnically-related aspects that could present risks to a project, and the need to address such items in a GBR. The Risk Register and GBR are complementary. The Risk Register process identifies, among other items, key geotechnical, construction, and third party risks. The GBR has the opportunity to provide a contractual platform for describing how certain risks have been addressed in the planning and design, and how other risks are to be allocated and managed during construction.

While the use of Risk Registers is strongly encouraged during the design of underground construction projects, the presentation and discussion of the details of a Risk Register evaluation process are best contained in other documents, perhaps Geotechnical Memoranda for Design such as discussed in Section 3.2.

Baselines by Tunnel Reach

When the tunnel passes through two or more distinct geologic regimes, it may be practical to define expected conditions in terms of separate baselines, and that is commonly done. It would follow that ground conditions and physical properties, as well as tunnel support measures, should be baselined within the context of these different regimes or reaches.

However, many tunnels cross geologic settings in which different ground conditions are completely intermixed in an unknown manner. For these instances, there is no rational basis for subdividing the tunnel into reaches with individual baselines. As an example, for most rock tunnel projects, a baseline would be given for expected maximum groundwater inflow for the entire tunnel in the absence of grouting for water cutoff. Whether or not this baseline was exceeded would not be known until the last day of excavation.

As another example, a tunnel might be expected to encounter only one geologic formation. The manner in which the various geotechnical parameters are distributed throughout the formation are not known, and cannot be reasonably predicted within the context of reach-by-reach baselines. For these instances, baselines defined for the entire tunnel drive are considered the more appropriate approach.

The Need for Qualified Preparers and Reviewers of GBRs

A number of comments have been made with regard to the importance of having suitably qualified geotechnical, design, and construction professionals involved in the preparation and review of GBRs. Experience has demonstrated that the ability to prepare a foundations report or a geotechnical report for design is insufficient background to prepare a GBR. Others have suggested that for projects large and small, Owners would be well served to institute a prequalification process at the preparer level, at the reviewer level, or both.

Another concern that has been raised is that given the recommended practice of developing the GBR at the later stages of design, insufficient review of the final GBR could result. It was considered that a minimum of two rounds of independent review of an advanced version of the GBR will improve clarity of content. Also, the value of a thorough page-by-page review of the GBR together with the other Contract Documents was underscored as a critical step to ensuring consistency and compatibility among the Contract Documents.

Use of GBRs on Small Projects

The extension of the GBR concept to areas of construction other than tunnels and shafts has received healthy debate, with the majority of the workshop participants acknowledging that more and more projects were utilizing this approach, and that the suggestions provided in Chapter 7 were both helpful and timely. Table 3 is offered knowing that improvements will be realized as the concept is implemented and further lessons are learned. It is hoped that foundation engineering practitioners and Contractors will advance a new guidelines publication more closely directed to this growing area of GBR application.

Baseline Measurement and Verification

This document stresses the importance of presenting GBR baselines in clear quantifiable and measurable terms. One comment raised by several workshop participants was that underground construction contracts should address the measurement and verification of baselines during construction as a matter of course. Guidance could be provided in the context of either specific activities to be accomplished from the inception of construction, or in the event that a dispute arose. Such guidance could be provided in the GBR or in the specifications.

It is considered that the measurement of certain aspects, e.g. groundwater inflows at the portal, shaft, or heading, can and should be provided for within the specifications. The documentation of boulders encountered during a soft ground tunneling project might also be accomplished with relative ease. However, a number of owners have resisted more extensive programs such as mandatory sampling and testing of rock conditions during the project to measure parameters such as compressive strength or abrasivity, due to the anticipated delay and costs of such programs. They offer that if the Contractor believes at some point in time that a DSC exists, the Owner will, upon proper notice, expend the necessary investigative costs at that time.

Certainly establishing a suitable process that assures that both parties will agree as to what the encountered conditions are is an important step toward avoiding and resolving disputes. For the parties to agree on how and when certain measurements are to be accomplished, based on guidance provided in the Contract Documents, bears consideration. However, such agreements might also be reached in the instance of an actual dispute, following the wisdom that "if it ain't broke, don't fix it." For example, in one recent rock TBM project, the Owner and Contractor agreed to have a mutually acceptable independent consultant carry out geologic mapping during the excavation. It was agreed that neither party would dispute the mapping results, and that the mapping costs would be shared. These agreements were made post-contract award.

While these types of agreements can and should be sought, they need to be implemented on a project-by-project, case-by-case basis. Inclusion of limited guidance provided in the Contract Documents should be helpful.

List of Abbreviations

DSC	Differing site condition
DRB	Disputes Resolution Board
DRBF	Disputes Resolution Board Foundation
EPB	Earth pressure balance
GBR	Geotechnical Baseline Report
GDR	Geotechnical Data Report
GIR	Geotechnical Interpretive Report
TBM	Tunnel boring machine
USNCTT	U.S. National Committee on Tunneling Technology
UTRC	Underground Technology Research Council

References

Better Contracting for Underground Construction (1974). U.S. National Committee on Tunneling Technology, National Academy of Sciences.

Better Management of Major Underground Construction Projects (1978). National Academy of Sciences.

Geotechnical Baseline Reports for Underground Construction (1997). Technical Committee on Contracting Practices of the Underground Technology Research Council, American Society of Civil Engineers.

Geotechnical Baseline Reports – A State of the Practice Review (2005), Essex, R.J. and Bartholomew, S.H., Rapid Excavation and Tunneling Conference, June, Seattle.

Geotechnical Site Investigations for Underground Projects (1984). National Academy of Sciences, Volumes I and II.

Levels of Geotechnical Input for DB Contracts for Tunnel Construction (2001). Robinson, R.A., Kucker, M.S., and Gildner, J.P., Rapid Excavation and Tunneling Conference, June, San Diego.

Recent Developments in the Use of Geotechnical Baseline Reports (2000), Essex, R.J. and Klein, S.J., North American Tunneling Conference, June, Boston.

The Differing Site Conditions Clause in Underground Construction (1985). Mathews, A.A.

Dispute Review Board Practice and Procedures Manual, Dispute Resolution Board Foundation, 2007.

Index

bidding 8, 38–39
budget allocation, contract document review 46

case reports 40–42
claims, minimizing 44
construction costs 16, 47
construction manager responsibilities 50
construction methods, relating to geotechnical baseline report 33–34; 35–36t
consultants 22, 26
contract clauses 13–15, 18, 54
contract documents 4–5, 8; consistency 28, 30; relation to geotechnical baseline reports 20–21; standard 13
contractor responsibilities 49–50

Deep Tunnel Sewerage Scheme, Singapore 40–41
delays 32
design team responsibilities 48–49
differing site conditions 37; contract clause 13–15, 18; federal clause 14, 54; modifications 15
dispute resolution 50, 52–53

excavations, construction considerations 25–26

foundations, construction considerations 25–26; design considerations 25
funding, effect on geotechnical baseline reports 43

geologic information, sources 23
geologic setting 23

geotechnical baseline reports, benefits 6–7; contents 22–27, 23–26t; data interpretation 5; effectiveness 51; enforcing 52; for bidding 38–39; for construction 39–40; for design-build contracting 38–40; goal of 16–17; items to be addressed 16–17; preparation 30–31, 51; purpose of 6; top issues 51–52; uncertainty 54–55; uses 5–6
geotechnical conditions, uncertainty 18–19, 53
geotechnical data reports, for design-build contracting 37–38; interpretation 9–10; role in risk identification 10; types of 11–12
ground characterization 24

highway improvement 42

Lake Hodges to Olivenhain Pipeline, San Diego 42

measurement 57–58

Niagara Tunnel Project, Ontario 41–42

owner participation in geotechnical baseline reports 31, 43–44
owner responsibilities 48
owner review, budget allocation 46 contract documents 46; geotechnical baseline reports 40

payment quantity provisions 54
pipelines 42; geotechnical baseline reports 57
project description 23
project overview 21

project size 32–33

reviewers, geotechnical baseline reports 40, 57
risk allocation 16, 27, 32, 44–45
risk factors, identifying 33–34
risk management 6, 45–46, 56

Sea to Sky Highway Improvement Project, British Columbia 42
Seattle Transit Program 41
sewage systems 40–41
shafts, construction considerations 25; design considerations 24–25
site conditions, contract clause 13–15, cost to owner 19–20
site exploration, for design-build contracting 37

small projects 32–33, 57
subsurface data, interpretation 15

terminology 27–28, 29*t*
Tren Urbano Subway, San Juan, Puerto Rico 40
tunnels 41–42, 56–57; construction considerations 25; design considerations 24–25

underground construction, reference documents 8

verification, geotechnical baseline reports 57–58